SERVICE-LEARNING
IN HIGHER EDUCATION

Service-Learning in Higher Education

Critical Issues and Directions

By
Dan W. Butin

SERVICE-LEARNING IN HIGHER EDUCATION

First published in 2005 by
PALGRAVE MACMILLAN™
175 Fifth Avenue, New York, N.Y. 10010 and
Houndmills, Basingstoke, Hampshire, England RG21 6XS
Companies and representatives throughout the world.

PALGRAVE MACMILLAN is the global academic imprint of the Palgrave Macmillan division of St. Martin's Press, LLC and of Palgrave Macmillan Ltd. Macmillan® is a registered trademark in the United States, United Kingdom and other countries. Palgrave is a registered trademark in the European Union and other countries.

ISBN 1–4039–6877–2

Library of Congress Cataloging-in-Publication Data is available from the Library of Congress.

A catalogue record for this book is available from the British Library.

Design by Newgen Imaging Systems (P) Ltd., Chennai, India.

First edition: July 2005

10 9 8 7 6 5 4 3 2 1

Transferred to digital printing in 2006

CONTENTS

Preface: Disturbing Normalizations of Service-Learning

Dan W. Butin

> . . . teaching is a question of strategy. That is perhaps the only place where we actually get any experience in strategy, although we talk a lot about it.
>
> Spivak 1989, p. 146

Service-learning appears ideally situated to make an impact in the classroom and in the world. Combining theory with practice, classrooms with communities, the cognitive with the affective, service-learning seemingly breaches the bifurcation of lofty academics with the lived reality of everyday life. Service-learning speaks to our sense of duty and fairness in the world: those who can supporting those who cannot, giving opportunities to those left behind.

These are grand narratives, in the best sense of the term, about what can be achieved through civic engagement, about attending to the plurality and diversity of the United States, about, simply, making a difference. And in many ways, service-learning has achieved just that. Almost one thousand postsecondary institutions are members of Campus Compact, the national coalition of college and university presidents committed to the civic purpose of higher education. Tens of thousands of public schools and millions of K-16 students engage in community service or service-learning each and every year. The human, fiscal, and institutional resources committed to the betterment of academic practice and community renewal is vast.

And yet. Such narratives must also be examined for their more troubling assumptions and implications: Who defines such narratives? In what terms? To what ends? For whose benefits? With what (unintended) consequences? This is not simply a cynical and relativistic appropriation of Lyotard's definition of the postmodern condition as the "incredulity to grand narratives." This is a fundamental grappling with the very heart and soul of service-learning theory and practice.

Service-learning in higher education is a potentially transformative pedagogical practice and theoretical orientation; it is, in Spivak's words, a question of strategy. By this I mean that service-learning challenges our static notions of teaching and learning, decenters our claim to the labels of "students" and "teachers," and exposes and explores the linkages between power, knowledge, and identity. Moreover, service-learning provokes by moving against the grain of traditional practice in higher education: it is a deeply engaging, local, and impactful practice.

This book examines the limits and possibilities of positioning service-learning as such a strategic practice. The contributors to this book explore and expand upon the tensions, troubles, and potentials of a service-learning that is dangerous to the educational status quo in higher education. By dangerous, I mean a pedagogical practice and theoretical orientation that provokes us to more carefully examine, rethink, and reenact the visions, policies, and practices of our classrooms and educational institutions. By dangerous, I mean a pedagogical practice and theoretical orientation that forces us—as faculty, administrators, and policymakers—to confront the assumptions under which we teach and learn and the implications for doing so in one way rather than another.

Service-learning makes us take a stand by acting up and acting out. It is much easier to teach within the boundaries of the normal. While I do not suggest that lecture-hall or seminar-style teaching is "easy," I do suggest that such teaching sidesteps and thus suppresses fundamental questions of higher education pedagogy: How is knowledge created and by whom? What is the "usefulness," if any, of disciplinary knowledge? What is the role of higher education in a liberal democracy? What is the role, moreover, of students, faculty, and institutions in their local and global communities? While the answers will obviously differ across institutions, the questions do not. Put otherwise, the normative silence on pedagogical practice by individual faculty and higher education institutions promotes and perpetuates traditional models of teaching and learning. This privileges top-down presumptions of knowledge transfer from faculty to students and power relations between institutions and community, and institutions and faculty. By implementing powerful service-learning programs, individuals must act up the institutional hierarchy.

Likewise, acting out—outside of traditional departments, outside of physical classroom walls, outside of the proximity and "safety" of the academic campus—is a dangerous endeavor. I do not mean the danger associated with coming into contact with some "exotic" or "threatening" Other outside of the bounds of the normal. Such a belief is grounded in the colonialist and racist mythologizing of those "below" as equivalent to the "primitive savage." I am instead referring to the danger for individual faculty on pragmatic, political, and existential grounds.

It is extremely difficult to pragmatically implement a powerful service-learning program. It takes foresight, time, organizational capabilities, creativity, networking skills, tolerance for ambiguity, willingness to cede sole control of classroom learning, and an acceptance of long-term rather than immediate increments of progress. It takes convincing—oneself and others—that the boundaries of academic disciplines, classroom walls, and institutional parameters are socially constructed and thus changeable. Yet, pragmatically speaking, such social constructions create normative pressures, solid walls, and clear institutional structures, all of which must be circumvented or worked through in order to implement service-learning programs.

Service-learning is also politically dangerous for individual faculty. It is a practice that might not be rewarded by traditional tenure and promotion guidelines, that questions (either implicitly or explicitly) colleagues' pedagogical practices, and that has the potential to turn out badly in a very public and glaring way. Given the high-stakes nature of the tenure review process, engaging in a nontraditional methodology is a disheartening proposition for new and junior faculty.

Finally, service-learning is an existentially dangerous endeavor. By this, I mean more than an individual's necessary fortitude and courage to confront and overcome the pragmatic and political obstacles to implementing service-learning. I mean that service-learning, when deeply done, subverts some of our most foundational assumptions of our sense of identity as higher education faculty. We must rethink the belief that academic knowledge comes directly from us, in a classroom, based on a written text, and assessed objectively. We must acknowledge our students as active, reflective, and resistant agents in their own educational processes. We must come to terms with the reality that our particular expertise may have very little currency (or even relevance) in the messy and complex world outside our classroom walls.

In many ways, therefore, this book works at the margins. It propounds a view and enactment of service-learning at the margins of traditional service-learning theory and practice. It offers a form of higher education pedagogy at the margins of academia. And it confronts the necessity for individual faculty to explore the margins of their own comfort zones. The contributors to this book do not ignore these issues. Rather, we willingly embrace this examination of the margins. For we are interested in provoking and providing a sustained, deliberate, and constructive examination of how service-learning *works*. I mean this in three distinct ways.

First, what actually occurs when we *do* service-learning? What are the micro-politics and micro-practices that are engaged in—by students, professors, and the individuals outside of higher education—when service-learning is enacted? How do our assumptions (e.g., of service, of assessment, of voice) inform how we develop and implement service-learning? How do our very use of space and words impact what we achieve? The contributors in this book examine the inner workings of how we do service-learning.

Second, how does service-learning *work* as a storyline and as a narrative? How does it shape participants' perspectives of what they do and why? What are the consequences of choosing one type of storyline rather than another? Choosing a storyline of "serving," for example, has vastly different assumptions and implications than choosing a storyline of "being with." We explore how positioning service-learning in different ways constructs and constrains what we can and cannot do through the experience.

Finally, what is it about service-learning that makes it *workable*? What are the diverse ways in which service-learning works to promote learning and change for all stakeholders in the process? What needs to be done—on individual, social, and institutional levels—to support the sustenance of a transformative service-learning practice?

The contributors to this book believe that this type of questioning is vital to the growth and vibrancy of the service-learning field. Without a constant questioning, there is the potential for self-serving; without a vision of what is possible comes the potential for just doing. I am not advocating for the wholesale overthrow of the "grand narratives" of the field. Not only is this naïve, but disregards how thoroughly service-learning has already become a part of the higher education landscape.

Instead, I suggest that service-learning truly be viewed as a question of pedagogical strategy. By this, I mean that it is a conscious intervention into local and highly complex contexts. Foucault (1997) suggested that one can never escape relations of power and "regimes of truth"; rather, "one escaped from a domination of truth not by playing a game that was totally different from the game of truth but by playing the same game differently, or playing another game, another hand, with other trump cards . . . by showing its consequences, by pointing out that there are other reasonable options" (pp. 295–296). Much like service-learning reveals the limits of traditional models of teaching and learning, we attempt to reveal how service-learning can be played differently, with different, as it were, "trump cards."

In order to do so, we must restate how service-learning is to be conceptualized. Let me thus begin by suggesting that service-learning is a culturally saturated, socially consequential, politically contested, and existentially defining experience.

Service-learning is culturally saturated to the same extent that any other complexly enacted process is in our society. It bears the assumptions and implications of our cultural models of, for instance growth, progress, individualism, and agency. More forcefully, one cannot separate what we think from who we are. The implications are that service-learning must be "read" as any other cultural practice, for there is no transparent, neutral, and objective position by which anyone and everyone understands what we mean when we say that we "do" service-learning.

Service-learning is socially consequential in the sense that all outcomes of the service-learning process, no matter how great or small, have impact in and on the world. For irrespective of the actual scope, duration, or outcome, all service-learning is done *by* individuals *with* other individuals. Whether this is explored through a social justice perspective (Freire 1994) or through a developmental identity framework (Baxter Magolda 1999; Tatum 1992), one must attend to the consequences of service-learning from the minutiae of individuals' interpretations to the vastness of an entire community's change.

Service-learning is politically contested in that it is fundamentally an attempt to reframe relations of power. This may be thought of in traditional models of empowering bottom-up change or in more postmodern notions of destabilizing foundational assumptions of knowledge, power, and identity. In either case, service-learning promotes a deviation from the status quo, if only because (at minimum) it just does things differently. On a deeper level, service-learning disturbs our society's penchant for security, order, and control, all of which are presumed to be synonymous to "safety." But as articulated earlier, service-learning is not safe. It is anything but safe. As such, all interventions that promote such disturbances—to the individual, the institution, or the community—are deemed political and thus contested.

Finally, service-learning is existentially defining because it forces individuals (students, faculty, and community partners) to take a stance. In so doing, individuals must (consciously or not) define themselves by the decisions they make or refuse to make. One cannot remain neutral when engaging in service-learning. Even the attempt to remain so positions oneself in a particular resistant identity. If we truly accept Geertz's (1973) melodic phrase that we are all ensconced within "multiple webs of meaning," then it becomes clear that service-learning pulls strongly at the strings that bind us and support us. And in that pulling and pushing, we must as existential beings decide in which direction we will be moved.

If service-learning can be conceptualized in this way—as a culturally saturated, socially consequential, politically contested, and existentially defining experience—then we can begin to talk about a different game, a different strategizing, that may be at play within the service-learning field. Such strategies do not directly address issues such as legitimization, best practices, or even social justice. For these are terms saturated in assumed meanings and contested outcomes. While we use such terms and often even abide by them, I suggest that we cannot take them for granted. If service-learning is to avoid becoming overly normalized, we must continuously question and disturb our assumptions, our terms, and our practices. It is in this spirit that the contributors attempt to think and act outside of the normal about how service-learning is to be. In so doing, such strategies disturb the normalizations in place within the service-learning field. Not in order to do away with them so much as, again following Foucault here, to demonstrate their consequences and how other options may play out. By working through other modes of service-learning in other ways, we hope to open up the field for further questioning and experimentation. But enough talking a lot about it; the following synopses describe how the contributors in this book actually begin to play out these perspectives.

CHAPTER OUTLINES

The book is split into three distinct sections. Section I—"The Micro-Politics and Micro-Practices of Service-Learning"—offers in-depth examinations of the workings of service-learning. Section II—"Transformative Models of Service-Learning Practice"—provides exemplary programs that consciously and conscientiously enact service-learning as centrally embedded in the teaching and learning process. Section III—"Reframing the Institutionalization of Service-Learning"—explores issues central to the future of service-learning in higher education.

In chapter 1, Susan Jones, Jen Gilbride-Brown, and Anna Gasiorski examine an "underside" of service-learning rarely discussed: student resistance. This situation—the seeming absurdity of students resisting "doing good"—forces us to confront one of the central dilemmas of service-learning. Namely, that service-learning is not just a "do good, feel good" enterprise that can easily be embraced by all. Using a "critical developmental framework" (by linking anti-oppressive scholarship with developmental psychology's work on self-authorship [e.g., Kumashiro 2002; Baxter Magolda 1999]), Jones, Gilbride-Brown, and Gasiorski elegantly show us the different paths and reasons for student resistance. Moreover, they begin to map out how to engage such resistance through "consciousness bridging" for both students and faculty around the contentious issues engaged through service-learning.

This is an extremely valuable beginning for the book, for it underscores the depth and rigor with which we must attend to the lacunas, the inconsistencies not often apparent when we just "do" service-learning. Put otherwise, our students' "not getting it" is not simply an oversight, on their part or ours. It is, again, a strategic move by students (and by faculty, in not attending to this "not getting it"), that is grounded in certain assumptions and with certain consequences. Jones, Gilbride-Brown, and Gasiorski insightfully reframe student resistance within service-learning courses and, in so doing, set the stage for the chapters to come.

In chapter 2, Raji Swaminathan investigates one of the central themes of service-learning: listening to the voice of the "served." Through a multiyear ethnography at an urban public high school, Swaminathan details how service-learning, no matter how well-meaning and how carefully planned, founders without students' voices. She shows how a high school set up with service at the very heart of its mission and pedagogical practice nevertheless must confront that students explicitly, systematically, and with immense consequences, begin to question and, ultimately, change what is meant by service and learning to themselves and to the community. Chaos theory is premised on the reality that miniscule changes in starting conditions have profound impacts on subsequent outcomes; likewise, Swaminathan's chapter explicates the issues surrounding whose voices are heard (and whose are not) when service-learning is set up, and the implications for such selective hearing.

In chapter 3, Sue Ellen Henry examines another central issue within service-learning: the implicit server–served binary. In a theoretically imaginative move, Henry explores how first-generation college students understand their service-learning experience and, by extension, themselves. At issue is that first-generation college students are oftentimes "serving" the very population they were themselves a part of just one or two years earlier. As Henry asks, "How do students who occupy both privileged and underprivileged status understand themselves and their multiple identity categories . . .?" Her findings shed light on the impact of class on service-learning practice and offer a deeper understanding of the predicament of first-generation college students—more and more of whom are to be found within today's higher education institutions. Finally, and I believe most importantly, her chapter begins to work through the deep assumptions and implications of the server–served binary upon service-learning theory and practice.

Tiffany Dacheux, in chapter 4, offers a response to and extension of Henry's chapter. Dacheux provides a voice for this process of liminality; hers is an exploration of being a part of both and neither worlds—college and community—at the same time. Dacheux challenges us to question the neutrality and "progressive" nature of becoming educated. Education, she suggests, changes us by moving us away from our past; it is in many ways an archetypal journey in becoming different. This is especially the case for those college students who do not arrive on campus with the requisite cultural and social capital presumed by privileged, white, middle-class norms. Her perspective thus importantly reveals the problematic potential for service-learning to reify those being served, and, more positively, the transformative potential for service-learning to support first-generation college students' becoming a part of a particular community.

In chapter 5, Caroline Clark and Morris Young rethink what it means to "change places" in service-learning. They point out that seeing from another person's point of view is neither so easy to accomplish nor so simple to talk about. Through a rigorous theoretical framework, Clark and Young argue that there are three types of spaces that we must attend to in service-learning— perceived, conceived, and lived space. Clark and Young demonstrate how service-learning scholars and practitioners can better theorize their practice by attending to power relations among individuals (perceived space), by taking advantage of "in between" spaces (made possible by conceived space), and by carefully understanding the actual positioning of individuals within their service-learning environments (lived space). This is a critical development for service-learning because it clearly and rigorously begins to explicate how to better think of service-learning as a relational, interactional, and complexly lived practice.

In the final chapter of this section, I work through the notion of service-learning as a paradigmatic example of postmodern pedagogy. I do so through the seemingly heretical embrace of Stanley Fish's demand that higher education reject the liberal political positions ("diversity," "civic engagement") that are also at the heart of traditional conceptions of service-learning. Specifically, I enumerate how service-learning acts to confound students' desire for closure. I suggest, again following Fish, that service-learning can be viewed as a "self-consuming" pedagogy that avoids (if we open ourselves to it) the reductionism and essentializing of finding "the" answer; that instead, service-learning works to disturb students' notions of static truth in order to allow for a more careful "reading" and analysis of the experience.

The next section of the book provides five distinctive service-learning practices. These are, to my mind, truly exemplary models that embrace service-learning at the heart of their theoretical and pedagogical practice. It is important to note that these models are not normative in the traditional sense used within the service-learning field. They do not advocate a certain number of on-site hours; they do not propound the best way to promote reflection; they do not have uniform or even compatible outcome measures. What makes them normative is their insistence that service-learning be thought of *as the course*, rather than as an add-on to a course. Put otherwise, without the service-learning component there is no course.

In chapter 7, Jordi Comas, Tammy Bunn Hiller, and John Miller describe an introductory management course taught at Bucknell University. Distinctive to this course is that students are assessed on a "double bottom line": doing service in the community funded through their own start-up operations. Working in teams throughout the semester, students develop practices of not just efficient and effective managers, but as invested-in-the-community managers. The course is grounded in three convictions that understanding how institutions function and managers manage is an essential general knowledge, that complex managerial problems necessitate active cooperative learning, and that people who will inevitably bear managerial responsibility and exercise formal authority must learn to think broadly and critically about their roles in society. Through a "communities of practice" theoretical framework (Wenger 1998), Comas, Hiller, and Miller offer an important real-world alternative to the instrumental model of management.

In chapter 8, Susan Dicklitch provides a vivid example of the power for linking service and learning in a political science course. Dicklitch constructed a course—Human Rights–Human Wrongs—that explored issues of human rights in general and U.S. asylum policies in particular. Students serve as researchers for community partner organizations on asylum seekers' cases at York County Prison (PA), the second largest detention center in the United States for asylum cases to be decided by the Bureau of Citizenship and Immigration Services and the Department of Homeland Security. Through interviews with the detainees and intensive research on immigration policies, human rights theory, case law, and the specific situations of each asylum seeker's story and country of origin, students create culminating immigration court-ready documents and legal briefs for the detainees. The impact on students and on the asylum seekers has been profound; most vividly, one of the detainees who was granted asylum came and spoke to Dicklitch's class upon being released: "Without the student's help, I would not be standing in front of you now, free, telling you my story."

In chapter 9, Marilynne Boyle-Baise and Paul Binford describe how one high school class unearthed and made public the historical stories and data of a once-segregated local school. Using a framework of multicultural service-learning (Boyle-Baise 2002), Boyle-Biase and Binford demonstrate how students in this semester-long project pondered questions of past and present racism, of how history is constructed, and of how they viewed themselves and their actions as citizens. This chapter provides an important example of how pedagogy and theory, academics and social justice can be blended together.

In chapter 10, Carl Milofsky and William Flack describe the organization, process, and outcomes of a field-based course for Bucknell University students in Northern Ireland. Students are placed with community organizations for an intensive three-week internship while concurrently taking an academic seminar on the history, politics, and impact of the 30-year sectarian conflict between Protestants and Catholics. There is a complete melding of academic and community learning as students learn about and experience the tensions and resolutions between individuals and groups in a context that is of paramount political and personal importance to their hosts. The students and professors vividly come to understand how the "personal is political." They also provide the opportunity for sustained discussion across individual and political barriers through a structured public panel discussion that brings together Bucknell students and local members and organizations to discuss the relations of the community and thus "to make local affairs visible and important to a wider audience."

In the final chapter of this section, Lori Pompa provides a detailed portrait of the "Inside-Out Prison Exchange Program," a set of courses she developed and teaches at Temple University and at the Graterford Maximum Security Prison outside of Philadelphia. The Inside-Out program is based on the concept of teaching college-level coursework within a correctional facility to both "inside" (incarcerated) and "outside" (undergraduate) students. By linking the program description, theoretical perspectives on such "border crossings," and the voices of the inside and outside students, Pompa disrupts our boundaries of who teaches whom. She, moreover, forces us to consider the profound realization that "service" may have much more to do with "being with" someone rather than "working for" someone.

The last section of the book presents two chapters that I view as critical to engage with in order to more fully think through future directions for service-learning scholarship and practice. In chapter 12, James Birge explores the "aesthetic basis" for service-learning. Birge's argument is that we must "consider a deeper foundation to service-learning . . . [through] the *connection* between service-learning practice and who we are as individuals and how we choose to act in the world." Birge suggests that as the service-learning field grew and focused upon its own legitimization and institutionalization, it failed to hold onto and examine the deep existential questions of why we as students, faculty, and administrators looked to service-learning theory and practice in the first place. Birge gives due credit to the advances and successes made possible by the "pragmatic elements" of service-learning practice. Yet, he notes, "without addressing the aesthetical underpinnings to our practice of service-learning, we may be building the structure of service-learning that lacks a deeper connection to the fundamental reasons for the work, and ultimately disables the foundation and sustainability of service-learning." No one that I am aware of has thoroughly begun to examine this issue; Birge does so with sensitivity and respect, and I think it opens the door for continued and necessary investigation.

Finally, in chapter 13, Matthew Hartley, Ira Harkavay, and Lee Benson provide a critical addition to the institutionalization literature within service-learning. Namely, they report on a study they conducted of four distinctive institutions (Swarthmore College, Tufts University, the University of Pennsylvania, and Widener University) at diverse stages of institutionalizing service-learning. Drawing from organizational theory (e.g., Goodman and Dean 1982), Hartley, Harkavay, and Benson explicate the diverse factors that influence degrees of institutionalization and levels of institutional commitment to service-learning. This is crucial as colleges and universities attempt to implement policies and practices thorough and wide-ranging enough to truly institutionalize service-learning on their campuses. Hartley, Harkavay, and Benson's work is therefore vital; it suggests possibilities for strategic interventions and offers a clearer perspective of the inevitable hurdles faced by higher education institutions across the country grappling with institutionalizing service-learning.

DISTURBING NORMATIVE FOUNDATIONS

This book revolves around three primary strands—theoretical, definitional, and pedagogical—that I believe are critical to examine for future discussions within the service-learning field. The first strand is a renewed critical investigation of the founding theoretical assumptions within service-learning. The second strand is an attempted redefinition of service-learning. The third and final strand is a reclaiming of service-learning as a truly transformative pedagogical strategy. Let me take each in turn to suggest how this book offers future directions to service-learning scholars and practitioners by disturbing the implicit normative boundaries of what we take to be service-learning.

The first strand stems from numerous chapters' consistent questioning of the founding theoretical assumptions for doing service-learning. This is seen in some students' resistance to service-learning, to the implicit dichotomization occurring in the server–served binary, and in the tension of whose voices are heard and privileged (see also Butin 2005). While each of these issues may be viewed as a technical problem fixable through methodological modifications (e.g., better pre-service orientations, greater participant buy-in), I suggest instead that such issues arise due to the inadequately articulated foundational assumptions of service-learning.

It may be fruitful to view an analogous field for insights into the tensions within the service-learning community. Multicultural education has spent the last 30 years grappling with and developing the distinctions it is premised on (Banks 1996; Sleeter and Grant 2003; see also Butin 2003b). The civil rights movements of the 1960s and 1970s shattered the ameleoristic presumptions of a paternalistic and assimilationist "melting pot" view of educational practices. Instead, multicultural educators developed a host of critical theories premised on fundamentally different assumptions of what counts as multicultural education: ethnic studies perspectives focused on educational practices and norms aligned to the cultural perspectives of those being taught (e.g., queer studies, women's studies, Afrocentric schooling [Asante 1998; Rich 1979]); difference multiculturalists emphasized the diversity of means by which we could think of learning, intelligence, and success (e.g., multiple intelligence [Gardner 1983]); critical multiculturalists demanded that issues of equity and social justice be at the heart of educational practices (e.g., detracking, problem-posing education [Freire 1994; Oakes 1985]).

Multicultural educators have thus put forward multiple alternative articulations in attempts to rethink and reframe what multicultural education should be premised on and moving toward. Every theoretical strand can be seen as a specific response to specific pedagogical or political problems (e.g., continued lack of equitable outcomes; lack of cultural congruence between students and textbooks). What is glaring in the service-learning field vis-à-vis multicultural education is the lack of analogous articulations of distinctive foundational assumptions. My point for the moment is not to critique the specific foundations of service-learning so much as point out the strong reliance on cultural/political conceptualizations of service-learning (see Butin 2003a, this volume). Such overreliance, I suggest, discourages a rigorous analysis and critique of the foundational terminology of service-learning for fear of a loss of meaning. Yet without multiple (and competing) foundational premises, the field is beholden to embracing potentially pragmatically limited and theoretically problematic articulations of service-learning.

This leads into a second strand of this book. The five distinctive service-learning programs described in the second section of this book highlight the definitional ambiguity and imprecision of the "service-learning" moniker: the service-learning rubric can be placed on a ten-hour, optional component to a course or on the models described in each chapter that demonstrate the depth, rigor, and power of a well-constructed service-learning program. Morton (1995) defined this as the difference between service-learning done in "thick" or "thin" ways, where "thick" service-learning programs authentically engaged academic and community goals whether done through charity, project-oriented, or social justice orientations.

These chapters clarify that "thickly" done service-learning programs have a host of attendant "family resemblances" and implications for practice. Namely, each chapter demonstrates the force of student immersion into specific local contexts, relevant and consequential learning outcomes, and profound respect for and attention to the process engaged in by all participants in the service-learning encounter. These characteristics, it might be noted, are highly resonant with Dewey's (1938) notions of experience and education. While it is beyond the scope of this section to delve into these commonalities, Dewey's (1959) demand that "education must be conceived as a continuing reconstruction of experience; that the process and the goal of education are one and the same thing" (p. 27) is highly suggestive for the reconstitution of how we go about articulating service-learning. Namely, that the service-learning experience is at the heart of, *if not indistinguishable from*, an academic course, however that may occur. No service-learning, no course; means and ends become indistinguishable. This, I suggest, is an important political move, for issues of, for instance, legitimization, best practices, and social justice within service-learning become enfolded into the larger discussions of how such issues play out in pedagogy in higher education more generally.

The third and final strand, I suggest, is a reclaiming of service-learning as a truly transformative pedagogical strategy. As stated at the beginning of this chapter, I see service-learning as a culturally saturated, socially consequential, politically contested, and existentially defining experience. This book attempts to link micro- and macro-levels of analysis of the service-learning experience by suggesting that what we do as individual faculty is ultimately structured by institutional parameters. This basic sociological point in turn offers guidance for the institutionalization of service-learning in higher education. Put simply, it is incumbent to change our unit of analysis from the classroom to the institution (or even the local geographic community) in order to begin to see the power of service-learning to transform.

This book is meant as the beginning of a what will hopefully be a sustained discussion within and outside of the service-learning field. It is a discussion I believe critical to have at this historical juncture. Service-learning has, as mentioned, become an accepted if not assumed part of higher education practice. Yet its position is still tenuous if we expect more than a pedagogical, curricular, or institutional add-on. At one level, all of us will take what we can get; we live in an imperfect world with multiple constraints and pressures on our time, energy, and resources. And yet, no matter how much we are pushed and pulled in different directions, what cannot be questioned (it can of course be ignored, suppressed, or misunderstood) is that service-learning is fundamentally a question of pedagogical strategy. Whether we like it or not, we engage in such strategizing each and every day we step into the classroom. I hope that the chapters in this book support your thinking through and rethinking these strategic moves. I also encourage you to visit my website (www.gettysburg.edu/~dbutin) where a host of resources are provided to support and extend these discussions and thoughts.

ACKNOWLEDGMENTS

This book was first conceived as part of a research project funded by the Pennsylvania Campus Compact. A group of scholars—myself, Jordi Comas, Sue Ellen Henry (both at Bucknell University), and Katherine McClelland (at Franklin and Marshall College)—met over the course of a semester to work through some of the foundational theories and assumptions of service-learning. One outcome of this was a paper I presented at a 2002 service-learning conference at Franklin & Marshall, and that later became an article (Butin 2003a). It was also at this conference that I heard Susan Dicklitch speak about a new course she had created—Human Rights–Human Wrongs.

I came to realize that the service-learning I had been grappling with was simply another mode of my larger project of theorizing about and developing some of the limits and possibilities for rethinking how one does pedagogy. This book is my attempt, with the help of a wide range of scholars, to better understand the pedagogical and social justice limits and potentials of service-learning.

I am thus thankful to a number of individuals who assisted my progress down this path. Julie Reed, at the Center for Public Service at Gettysburg College (and who is now at the University of San Francisco) was a constant support to my questions, requests, and ideas for service-learning at Gettysburg College. James Birge provided support both through the Pennsylvania Campus Compact and as an intermittent interlocutor. Kathleen Knight Abowitz, Eric Bredo, and Sue Ellen Henry, whether they knew it or not, gave needed and timely intellectual sustenance. Amanda Johnson at Palgrave has made this book possible on many levels. She contacted me, out of the blue, about two years ago and encouraged me to work through and submit to Palgrave some ideas I had been working on at that time. This was a wonderful realization to a new scholar; that someone might actually be interested in the work I do. She has now helped guide this book quickly and professionally through the editorial process, and I am deeply grateful for such support. Finally, my wife, as usual, has been my best critic, reader, and discussant.

REFERENCES

Asante, M. K. (1998). *The afrocentric idea* (Rev. and expanded). Philadelphia: Temple University Press.

Banks, J. A. (1996). *Multicultural education, transformative knowledge, and action: Historical and contemporary perspectives.* New York: Teachers College Press.

Baxter Magolda, M. B. (1999). *Creating contexts for learning and self-authorship: Constructive-developmental pedagogy.* Nashville, TN: Vanderbilt University Press.

Boyle-Baise, M. (2002). *Multicultural service learning: Educating teachers in diverse communities.* New York: Teachers College.

Butin, D. W. (2003a). Of what use is it? Multiple conceptualizations of service-learning in education. *Teachers College Record,* 105(9): 1674–1692.

Butin, D. W. (2003b). The limits of categorization: Re-reading multicultural education. *Educational Studies,* 34(1): 62–70.

Butin, D. W. (2005). Identity (re)construction and student resistance. In D. W. Butin (Ed.), *Teaching social foundations of education: Contexts, theories, and issues.* (pp. 109–126). Mahwah, NJ: Lawrence Erlbaum Associates.

Dewey, J. (1938). *Experience and education.* New York: The Macmillan Company.

Dewey, J. (1959). *Dewey on education.* New York: Teachers College, Columbia University.

Foucault, M. (1997). *Michel Foucault: Ethics, subjectivity, and truth.* Edited by Paul Rabinow. New York: The New Press.

Freire, Paulo. (1994). *Pedagogy of the Oppressed.* New York: Continuum.

Gardner, H. (1983). *Frames of mind: The theory of multiple intelligences.* New York: Basic Books.

Geertz, C. (1973). *The interpretation of cultures.* New York: Basic Books.

Goodman, P. S. & Dean, J. W. (1982). Creating long-term organizational change. In P. S. Goodman (Ed.), *Change in organizations.* San Francisco, CA: Jossey-Bass.

Kumashiro, K. K. (2002). Against repetition: Addressing resistance to anti-oppressive change in the practices of learning, teaching, supervising, and researching. *Harvard Educational Review*, 72: 67–92.

Morton, Keith. (1995). The irony of service: Charity, project, and social change in service-learning. *The Michigan Journal of Community Service-Learning*, 2: 19–32.

Oakes, J. (1985). *Keeping track: How schools structure inequality*. New Haven, CT: Yale University Press.

Rich, A. C. (1979). *On lies, secrets, and silence: Selected prose, 1966–1978* (1st Edition). New York: Norton.

Spivak, Gayatri, with Ellen Rooney (1989). In a word. Interview. *Differences*, 1(2): 124–156.

Sleeter, C. E. & Grant, C. A. (2003). *Making choices for multicultural education: Five approaches to race, class, and gender* (4th Edition). New York: J. Wiley & Sons.

Tatum, B. D. (1992). Talking about race, learning about racism: The application of racial identity development theory in the classroom. *Harvard Educational Review*, 62(1): 1–24.

Wenger, E. (1998). *Communities of practice: Learning, meaning and identity*. Cambridge, UK: Cambridge University Press.

Notes on Contributors

Lee Benson, *University of Pennsylvania* Lee Benson is Professor Emeritus of History and Distinguished Fellow at the Center for Community Partnerships at the University of Pennsylvania. He collaborates with Dr. Ira Harkavy on action-oriented research and teaching designed to exemplify their conviction that the primary mission of the American university is to help realize an optimal democratic society and world. A nationally recognized leader in shaping the foundation of academic-based community service, Dr. Benson is co-executive editor of Universities and Community Schools and author of six books. He is currently revising a book-length manuscript with Ira Harkavy and John Puckett for publication titled *Progressing Beyond John Dewey: Developing and Implementing Practical Means to Realize Dewey's Utopian Ends.*

Paul E. Binford, *Jackson Creek Middle School & Indiana University-Bloomington* Paul Binford is a social studies teacher at Jackson Creek Middle School. He is a Lilly Endowment Fellow, and has published a simulation entitled, "Lincoln's Cabinet and the Sumter Crisis: a Simulation." As a doctoral candidate at Indiana University, his research interests include service-learning, citizenship education, and the history of social studies education.

James Birge, *Pennsylvania Campus Compact* James Birge is the Executive Director of the Pennsylvania Campus Compact, a coalition of college and university presidents committed to the civic purposes of higher education. Prior to working for Campus Compact, Dr. Birge was the Coordinator of the Center for Service Learning at Regis University in Denver, Colorado from 1992 to 1997. Dr. Birge's current scholarly interests focus on partnership development between higher education institutions, public scholarship, and the role of academic presidents as public leaders. He coauthored with Brooke Beaird and Jan Torres the chapter "Partnerships Among Colleges and Universities for Service-Learning" that appeared in *Building Partnerships for Service-Learning* (Jacoby 2003) and "Beyond the Shadow of the Ivory Tower: Higher Education and Democracy" that appeared in *A Blueprint for Public Scholarship at Penn State* (Cohen and Yapa, Eds. 2003).

Marilynne Boyle-Baise, *Indiana University-Bloomington* Marilynne Boyle-Baise is Associate Professor in the Department of Curriculum and Instruction at Indiana University-Bloomington. She teaches curriculum studies, social studies, and multicultural education. She serves on the Board of Directors for the National Council for the Social Studies. She has served on editorial boards for Theory and Research in Social Education and the International Social Studies Forum. Her current research focuses on multicultural service-learning, citizenship education, and community-based teacher education. She has published widely, and she has written a book, published by Teachers College Press entitled *Multicultural Service Learning: Educating Teachers in Diverse Communities.*

Dan W. Butin, *Gettysburg College*   Dan W. Butin is Assistant Professor of Education at Gettysburg College. He specializes in the social and cultural contexts of education, with an emphasis on issues of multicultural education and the sociology of education. His recent research and publications focus on the intersections of critical multiculturalism, poststructuralist thought, and alternative pedagogical strategies. Dr. Butin is on the editorial board of *Educational Studies.* His recent publications have appeared in *Teachers College Record* and *Educational Researcher* and he is the editor of a forthcoming book, *Teaching Social Foundations of Education: Contexts, Theories, and Issues* (Lawrence Erlbaum Associates).

Caroline Clark, *Ohio State University*   Caroline Clark is Associate Professor of Language, Literacy, and Culture in the School of Teaching and Learning where she coordinates the Drama, Language Arts, Literature, and Reading program and teaches courses in English Education. Her teaching and scholarship focus on adolescent literacy learning across formal/school and informal settings, and she has published work in the *American Educational Research Journal, Teachers College Record*, the *Journal of Literacy Research*, and the *Journal of Adolescent and Adult Literacy.* She has taught English language arts in both secondary and elementary settings and has worked with literacy learners in a variety of after-school programs.

Jordi Comas, *IESE/Universidad de Navarra and Bucknell University*   Jordi Comas is a visiting Assistant Professor of Management at Bucknell University, a doctoral candidate at IESE, and an independent consultant and educator. His interests include the sociology of organizations and education, especially the relationship between social networks and knowledge and learning. He has conducted research and presented internationally on the social status systems of high schools, experiential learning, school finance reform, the production and diffusion of management knowledge, and social networks.

Tiffany Dacheux, *Gettysburg College*   Tiffany is a student at Gettysburg College. She is currently pursuing a bachelor's degree in history and a certification to teach secondary social studies, both of which she expects to obtain in 2005.

Susan Dicklitch, *Franklin & Marshall College*   Susan Dicklitch is Associate Professor of Government at Franklin & Marshall College. She has published several articles on civil society and human rights in *Human Rights Quarterly*, *International Politics*, *International Journal*, *Journal of Contemporary African Studies*, and is the author of *The Elusive Promise of NGOs in Africa: Lessons from Uganda* (Palgrave). She teaches African Politics, The Politics of Developing Areas, and Human Rights.

William F. Flack, Jr., *Bucknell University*   Bill Flack is Assistant Professor of Psychology at Bucknell University, and a licensed clinical research psychologist. He edited *What is Schizophrenia?* with Daniel R. Miller and Morton Wiener (New York: Springer-Verlag, 1991), and *Emotions in Psychopathology: Theory and Research* with James D. Laird (New York: Oxford University Press, 1998). He was a post-doctoral fellow in clinical research in the Behavioral Science Division of the National Center for Posttraumatic Stress Disorder at the Boston Department of Veterans Affairs Medical Center from 1996 to 1998. He conducts research on the self-perception of emotion, psychological trauma, and interpersonal conflict. He is a codeveloper of the inter-university Consortium for Northern Ireland Studies, and of Bucknell's service-learning program in collaboration with the University of Ulster, Magee College, in L'Derry, Northern Ireland.

Anna Gasiorski, *Ohio State University*   Anna Gasiorski is a doctoral student in the higher education program in the School of Educational Policy and Leadership at the Ohio State University. Prior to her work at Ohio State, she received her Master's degree in Secondary Education from the University of Michigan and spent three years teaching high school Spanish. Currently, she is the instructor of an undergraduate service-learning course entitled *Leadership in Community Service*. As a doctoral student at Ohio State, her research interests include service-learning, student resistance, and university/community partnerships.

Jennifer Gilbride-Brown, *Ohio State University*   Jen Gilbride-Brown is a doctoral student in the higher education program in the School of Educational Policy and Leadership at the Ohio State University. She received her Master's degree in Higher Education Administration from the College of William and Mary in Virginia. For the past three years, she instructed an undergraduate service-learning course entitled *Leadership in Community Service*. As a doctoral student at Ohio State, her dissertation work involves an examination of white college students' understandings and perceptions of whiteness in the context of service-learning.

Ira Harkavy, *University of Pennsylvania*   Ira Harkavy is Associate Vice President and Founding Director of the Center for Community Partnerships at the University of Pennsylvania. Dr. Harkavy is an historian with extensive experience building university–community–school partnerships. He is executive editor of *Universities and Community Schools* and a member of the editorial boards of *Non-Profit Voluntary Sector Quarterly* and *Michigan Journal of*

Community Service Learning. Dr. Harkavy served as consultant to the U.S. Department of Housing and Urban Development to help create its Office of University Partnerships. He is the recipient of Campus Compact's Thomas Ehrlich Faculty Award for Service Learning; and under his directorship the Center for Community Partnerships received the inaugural William T. Grant Foundation Youth Development Prize sponsored in collaboration with the National Academy of Sciences' Board on Children, Youth and Families and a Best Practices/Outstanding Achievement Award from HUD's Office of Policy Development and Research.

Matthew Hartley, *University of Pennsylvania*   Matthew Hartley is Assistant Professor of Education at the University of Pennsylvania's graduate school of education. His research focuses on issues of organizational change and governance at colleges and universities. He is particularly interested in the role that mission plays in shaping institutional programs and policies. He has contributed a chapter (co-written with Elizabeth Hollander) to *Building Partnerships in Service Learning and Civic Responsibility in Higher Education.* A book based on Dr. Hartley's dissertation research, *A Call to Purpose: Mission-Centered Change at Three Liberal Arts Colleges,* was published in 2002.

Sue Ellen Henry, *Bucknell University*   Sue Ellen Henry is Assistant Professor of Education at Bucknell University. Her research interests include moral education, multicultural and democratic education, sociology and philosophy of education. Dr. Henry teaches several courses that include significant service learning requirements, such as Multiculturalism and Education and Social Foundations of Education. She has published in *Teachers College Record, Educational Theory, Educational Studies*, and the *Journal of Negro Education.* She is also on the editorial board for *Educational Studies*, and serves as a reviewer for *Educational Theory* and *The Journal of College Student Retention.* Before joining the faculty at Bucknell, she held a number of student affairs positions including Assistant Director of Residential Life at LaSalle University in Philadelphia and Program Coordinator for the Residential Colleges at Bucknell.

Tammy Bunn Hiller, *Bucknell University*   Tammy Bunn Hiller is Associate Professor of Management at Bucknell University. For the last ten years, she has engaged students in service-learning through her teaching in Management 101. In addition, she has recently developed an interdisciplinary service-learning course on organizing for justice and social change, aimed at developing students' collaborative capabilities as justice-oriented citizens. Her scholarship focuses on experiential, collaborative learning pedagogies, reflective practices, and case study research. She has presented numerous workshops on these issues at *Organizational Behavior Teaching Society* and *Academy of Management* conferences, and has published in *Journal of Management Education, Business Case Journal*, and *Case Research Journal.*

Susan Jones, *Ohio State University* Susan Jones is Associate Professor in the School of Educational Policy and Leadership and Director of the Student Personnel Assistantship Program at The Ohio State University. Before joining the faculty at Ohio State, she held a number of student affairs positions including Dean of students at Trinity College of Vermont and Assistant Director of campus programs and the Stamp Student Union at the University of Maryland. Dr. Jones's research interests and publications focus on identity development of college students, service-learning, qualitative methodology, and student affairs administration. She has been a two-term editorial board member of the *Journal of College Student Development*. She has twice served as an Ohio Campus Compact faculty fellow, was the recipient of a Scholarship of Engagement grant, and currently serves on the executive board of Ohio Campus Compact. She is a previous board member of the National Association for Women in Education and Chair of the Board of Trustees of Project OpenHand, an AIDS service organization in Columbus, Ohio.

John A. Miller, *Bucknell University* John A. Miller, Lindback Professor of Management at Bucknell University, also taught at Yale, INSEAD and Rochester, and in visiting professorships at Doshisha University in Kyoto, Japan, and in France. He established MG 101, an experiential management service-learning course, in 1979; versions of MG 101 are offered in more than a dozen colleges and universities. His teaching, scholarly, and consulting interests center on experiential learning methods, service-learning, and on the special needs of undergraduate general education students of management. Prof. Miller served as a member of the Board of Directors and as Executive Director for the Organizational Behavior Teaching Society. The Carnegie Foundation for the Advancement of Teaching has named Prof. Miller a Fellow of the Carnegie Teaching Academy.

Carl Milofsky, *Bucknell University* Carl Milofsky is Professor of Sociology at Bucknell University. He completed a post doctoral fellowship in Education at the University of Chicago in 1978. He has used field-based instruction in his courses since 1976 and has written a variety of articles about this sort of teaching as well as about university community collaborations in research, action, and teaching. His dissertation research was on the sociology of special education and he is author of *Special Education: A Sociological Study of California Programs* (1976) and *Testers and Testing: The Sociology of School Psychology* (1988). He edited the *Nonprofit and Voluntary Sector Quarterly* from 1990 to 1996. Having taught at Yale University from 1978 to 1982, he was a founding member of the Program on Nonprofit Organizations working primarily in the area of community and small nonprofit organizations. He has continued as a participant in PONPO research programs through the 1990s. He edited *Community Organizations: Studies in Resource Mobilization and Exchange* (New York: Oxford University Press,

1988) and *Community Chest* by John Seeley et al. (New Brunswick, NJ: Transaction Books, 1989).

Lori Pompa, *Temple University*   Lori Pompa has a Special Appointment Faculty position in the Department of Criminal Justice at Temple University, and is Founder and Director of The Inside-Out Prison Exchange Program, creating opportunities for social change through dialogue between those outside and those inside of our nation's correctional facilities. As a Soros Justice Senior Fellow, she collaborated with others on both sides of the wall to develop Inside-Out into a model for national replication. To date, 33 instructors from more than a dozen states have been trained in the Inside-Out approach. The Inside-Out National Instructor Training Institute conducts three week-long trainings each year. Employing experiential and service-learning pedagogies in her classes, she has taken more than 7,500 students into area correctional facilities in the past 12 years, and has worked with incarcerated men and women since 1985.

Raji Swaminathan, *University of Wisconsin-Milwaukee*   Raji Swaminathan is Assistant Professor of Education in the Department of Educational Policy at the University of Wisconsin-Milwaukee. She teaches courses related to urban education and alternative education. Her research interests are in the areas of youth and schooling, gender and schooling, and immigrant education. She has taught and conducted research at alternative schools working in India and England. She received her Ph.D. in cultural foundations of education at Syracuse University. Her articles have been published in *High School Journal* and the *Journal of School Effectiveness and School Improvement*.

Morris Young, *Miami University, Ohio*   Morris Young is Associate Professor of English and Director of Graduate Studies at Miami University in Oxford, Ohio. He received his doctorate from the Joint Ph.D. Program in English and Education at The University of Michigan in 1997. He teaches undergraduate courses in writing, graduate courses in composition and rhetoric and literacy studies, and courses in Asian Pacific American literature. His work has appeared in *College English, The Journal of Basic Writing, Critical Theory and the Teaching of Literature: Politics, Curriculum, Pedagogy, The Literacy Connection*, and *Personal Effects: The Social Character of Scholarly Writing*. His book *Minor Re/Visions: Asian American Literacy Narratives as a Rhetoric of Citizenship* (Southern Illinois University Press, 2004) examines the rhetorical construction of American citizenship through the narrative writing of Asian Americans. These narratives often tell stories of transformation through education, the acquisition of literacy, and cultural assimilation/ resistance, and offer a revision to the American Story.

SECTION I

THE MICRO-POLITICS AND MICRO-PRACTICES OF SERVICE-LEARNING

Getting Inside the "Underside" of Service-Learning: Student Resistance and Possibilities

Susan Jones, Jen Gilbride-Brown, and Anna Gasiorski

"People come to the food pantry driving nice cars. If they can afford those cars, why do they need to get food for free?"

"What do you expect from kids whose parents don't care enough to read with them at night and help them with their homework, like mine did?"

"Why should I think about people with AIDS? I thought 'you got AIDS, don't bother me, go away, you're going to die.' I didn't care about them."

"When I walked into [homeless shelter] for the first time, my heart was ripped out of me and stomped on. I couldn't believe my eyes . . . I wasn't sure this was an environment I wanted to be volunteering in."

"I don't see why we have to do all these readings and papers. They don't seem to have anything to do with what I am doing at my service site."

Service-learning is often heralded as a pedagogical strategy with "transformative potential" (Jones 2002; Rosenberger 2000). However, as these illustrative quotations from undergraduate students enrolled in our service-learning courses suggest, not all students are immediately, or gracefully, transformed by their experiences. Further, students' abilities to engage with all aspects of their service-learning courses depend on the intersection of their own sociocultural backgrounds (which become very apparent in community service environments), developmental readiness for such learning to occur, and the privileging conditions that situate college students in community service organizations in the first place (Jones 2002).

Much of the discourse and research in service-learning focuses on the positive student outcomes associated with this educational strategy (e.g., Eyler and Giles 1999). We do not want to dispute the positive outcomes associated with service-learning, as we too have experienced the "transformative potential" of engaging students in this work. However, scant attention is paid to the "underside" of service-learning. By "underside," we mean the complexities that emerge when undergraduate students engage with ill-structured, complex social issues present in the community service settings typically associated with service-learning courses. In such settings, previously held assumptions, stereotypes, and privileges are uncovered. When the "underside" is exposed, student resistance often ensues as the service-learning experience makes claims on students for which they are not prepared to process (Jones 2002; Kegan 1994).

We want to explore in this chapter, the *inside* of the *underside* of service-learning so that student resistance might be better understood and reconceptualized as a site for transformative potential. We engage this discussion of student resistance and possibilities through the lenses of a cognitive developmental theory of self-authorship (Baxter Magolda 1999, 2000a,b; Kegan 1994) and critical whiteness (Fine et al. 1997; Frankenberg 1993; Giroux 1997; Tatum 1999) in the context of a service-learning course taught as a critical pedagogy (Giroux 2003; McLaren 2003a). We refer to this integration of theoretical frameworks as a *critical developmental lens*. We place this theoretical discourse in dialogue with student voices through the presentation of three profiles of student resistance. These profiles are derived from our experiences of teaching undergraduate students in service-learning courses and from analysis of student papers and reflections on their experiences. We draw heavily from the words of students themselves and then turn a critical developmental lens on an interpretation of these words, in an effort to make meaning of how students are negotiating their service-learning experiences.

Drawing less from the service-learning literature and more so from the critical theory domain, "the concept of resistance emphasizes that individuals are not simply acted upon by abstract 'structures' but negotiate, struggle, and create meaning of their own" (Weiler 1988, p. 21). While student resistance can certainly emerge as a problematic and disruptive phenomenon in the classroom (Butin 2005), its presence reflects student negotiation with a meaning making process that holds open the possibilities for engagement with the "dynamics of power and privilege in service-learning" (Rosenberger 2000, p. 24), an oft-noted goal of service-learning, yet one rarely grappled with at great depth. Sonia Nieto (2000) captures well the dynamic interplay from which resistance in service-learning emerges:

> One cannot help but notice, for instance, that the primary recipients of community service are those who society has deemed disadvantaged in some way, be it through their social class, race, ethnicity, ability, or any combination of these. Those who do community service at colleges and universities, on the other hand, are generally young people who have more advantages than those they are serving.   (pp. ix–x)

In our service-learning courses, we found that undergraduate students are confronted with their own privileges and positions of power, often for the first time. The complicated environments of community service sites situate students in a " 'borderlands,' where existing patterns of thought, relationship, and identity are called into question and juxtaposed with alternative ways of knowing and being" (Hayes and Cuban 1997, p. 75). The borderlands represent "sites both for critical analysis and as a potential source of experimentation, creativity, and possibility" (Giroux 1992, p. 34). The complex dynamics of student resistance, then, emerge in the intersections of negotiated student identities, encounters in the borderlands, and intentionally provoked classroom dialogues focused on power and privilege.

SETTING THE CONTEXT

A quarterly ten-week Leadership Theories course that incorporates service-learning provided the context for student experiences and our analysis. The course engaged students with the question of "leadership for what purpose?" through course readings, class discussions, small group reflection, and three hours/week of service at a local community organization. We designed the course with an intentional focus on social justice and to reflect the goals of multicultural education. As noted by O'Grady (2000):

> Without the theoretical underpinnings provided by multicultural education, service-learning can too easily reinforce oppressive outcomes. It can perpetuate racist, sexist, or classist assumptions about others and reinforce a colonialist mentality of superiority. This is a special danger for predominately White students engaging in service experiences in communities of color. (p. 12)

Much is written in the service-learning literature about the appropriate goals for service-learning activities. Typically, this discussion is framed using the language of *charity* or *social change* (e.g., Kahne and Westheimer 1996; Morton 1995; Rhoads 1997). In short, an emphasis on charity situates students as those providing service *to* those in need (e.g., those "less fortunate") and feeling good about "helping" others. By contrast, an approach grounded in social change places students in relationship *with* those with whom they are serving and emphasizes the connections between student service and the larger social issues around which the community service sites are organized. Morton (1995) added the possibility of engaging in service in either "thick" or "thin" ways, distinguishing between service experiences that exhibit integrity, respect, and some connection to larger social structures and those in which relationships are not present, commitment to social issues is not cultivated, and students engage in their community service in potentially disingenuous ways.

Our service-learning courses are rooted in a social change framework. We have cultivated and sustained relationships with our community service partners, several for over six years. The course is designed with an intentional focus on power, privilege, and social justice. Intended learning objectives included the development of an understanding of leadership and social issues from multiple perspectives, as well as a personal philosophy of leadership through critical analysis of social issues and community involvement. Meeting these objectives occurred through readings that included both leadership theories and narratives capturing some of the life situations students would encounter at their service site, large class lecture and discussion, small group reflections, and a minimum of three hours/week at one community service organization for the duration of the quarter. The small group reflections consisted of all those students working at the same service site (never more than 10 students per group) and were facilitated by a graduate student with preparation in the theory and practice of service-learning.

The community service organizations partnering with the course represented a diverse array of social issues including HIV/AIDS, hunger and homelessness, and literacy. Relationships with these organizations were cultivated and sustained because of the close match between academic learning objectives and the mission and activities of the organization. Because research has demonstrated that direct service is linked to greater critical consciousness and understanding of complex social issues (Eyler and Giles 1999; Jones and Hill 2001; Jones and Abes 2003; Rhoads 1997), students at most sites provided direct service (e.g., preparing and delivering nutritious meals to people living with AIDS; providing one-to-one tutoring for kids struggling in school; packaging and distributing bags of food to individuals accessing the services of a food pantry; playing with children spending the day at a homeless shelter). In addition, students often had an opportunity to work side-by-side with individuals who utilize the services provided by the organization or other volunteers who have more deeply personal reasons for being there (e.g., loss of a friend to AIDS).

Students who enrolled in this elective course represented a variety of ages and academic majors. Nearly 70 percent of the students enrolled in this class are female and 80 percent are white. Our focus here is on the white students in the class as they represented all of those students who demonstrated resistance. Based upon what they indicated during the first class session, students' reasons for enrolling appear to represent a vast array of motivations including: requirement for a scholars program, heard good things about it from a friend, getting credit for volunteering was an easy A, wanted to build on leadership skills and community service experience from high school, and thought it would look good on a resume. The rare student articulated a vision for himself/herself that included increasing self-knowledge and deepening his/her commitment to social justice issues, equity, and civic action, some of our intended outcomes for the course. This course design, and the student resistance profiled, was consistent across sections and instructors each time the class was taught.

Turning a Critical Developmental Lens on Ourselves

As three highly educated, white, middle- to upper-middle-class women, we recognize that we bring to our interpretations of student profiles of resistance the realities of our own privileged backgrounds. We are also deeply committed to teaching for social justice and are involved in this particular leadership theories course because we care about educating future leaders who will be more attuned to issues of inclusion, justice, and decency. In our beliefs about the role of education in creating a more just society we guard against "fuming" from our side of the bridge and exuding an arrogance of the self-proclaimed socially conscious.

We also acknowledge that teaching service-learning classes from a social justice perspective and wrestling with student resistance is uncomfortable for us. It is tempting at times to settle the resistance; after all, we have been taught (and evaluated) to "maintain order" and impart knowledge in the classroom. We also appreciate that to fully engage with student resistance (which we think is necessary for new knowledge to grow) we must learn to appreciate the "pedagogy of the unknowable" (Ellsworth 1994) and "teach paradoxically" (Kumashiro 2004). Our practice shows us this is no easy task.

THEORETICAL FRAMEWORKS

Exploring student resistance as a complex phenomenon necessitates drawing from several theoretical frameworks as well as the caution that students always remain larger than their categories (Perry 1978). Understanding student resistance as a process of struggle, negotiation, and meaning-making anchors our analysis in the literatures on self-authorship and critical whiteness. Earlier research on the developmental outcomes of service-learning pulls these constructs apart and examines each independently. Typically, domains of development (e.g., psychosocial, cognitive, moral) and the development of social identities (e.g., racial, ethnic, sexual) are treated as discrete domains (Evans et al. 1998). Rarely is the intersecting nature of these domains of development and the dynamics involved in such interaction examined, particularly sparked by the pedagogy of service-learning. One study addressed these connections and reported, "What is enduring about service-learning is the likelihood of increasing integration of these domains . . . Because students were introduced to and developed relationships with both individuals and social issues with which they were unfamiliar, previously held notions of self and other were disrupted, challenged, and reconstructed" (Jones and Abes 2004, p. 163).

Self-Authorship

Self-authorship is one theoretical framework that does address intersecting domains of development. First defined by Robert Kegan (1994) self-authorship is:

> an ideology, an internal identity, a *self-authorship* that can coordinate, integrate, act upon, or invent values, beliefs, convictions, generalizations, ideals, abstractions, interpersonal loyalties, and interpersonal states. [The person] is no

longer *authored* by them, [the person] *authors* them and thereby achieves a personal authority. (p. 185, emphasis in original)

Drawing on her longitudinal research following young adults, Baxter Magolda (1999) describes self-authorship as "an ability to construct knowledge in a contextual world, an ability to construct an internal identity separate from external influences, and an ability to engage in relationships without losing one's internal identity" (p. 12). The journey toward self-authorship depends upon the integration of cognitive complexity, interpersonal maturity, and intrapersonal (identity) development; and locates student development as a process of moving from more formulaic, externally derived understandings of self to "foundational" meaning-making characterized by internally generated values and beliefs (Baxter Magolda 1999). Self-authorship is prompted by experiences of cognitive dissonance (Baxter Magolda 1999). Service-learning provides an especially rich opportunity to promote students' development toward self-authorship because of the dissonance created between previously held conceptions of self and new experiences, reflection on this dissonance, and new learning that occurs as a result (Baxter Magolda 2000a; Jones and Abes 2004).

Kegan's (1994) conception of a "consciousness bridge" provides a metaphor and strategy for understanding the nature of the learning that occurs in service-learning classes, student resistance to this learning, and support for the journey toward self-authorship. Building on the work of William Perry, Kegan's idea of the consciousness bridge describes the nature of developmental transitions particularly when individuals are "in over their heads" and the meaning individuals make of such transitions. The traverse across the bridge is one of "developmental transformation, or the process by which the whole ('how I am') becomes gradually a part ('how I was') of a new whole ('how I am now')" (Kegan 1994, p. 43). Further, Kegan conceived of the bridge builder as having "equal respect for both ends, creating a firm foundation on both sides of the chasm students will traverse" (p. 278) and the bridge building as a process of co-creation between the bridge builder and students so that students would play a part in constructing a bridge that "they could choose to walk out on" (p. 279). Presumably, the journey across the consciousness bridge brings a student closer to a self-authored identity.

While the theoretical framework of self-authorship provides an analytic lens for understanding how students are making meaning of their service-learning experiences, it does not squarely focus attention on the complexities that emerge when students confront issues of power and privilege. A potential downside to analyzing student experiences solely through a cognitive developmental lens is that it centers students as the object of scrutiny and analysis and results in a comment such as "this student just didn't get it" (Butin 2005; Jones 2002). The perspective of critical whiteness helps to situate these individual student voices in a larger social context, both in terms of the complicated environments in which service takes place and in addressing the issues of power and white privilege central to many service-learning courses.

Critical Whiteness

Critical whiteness is a theoretical framework focused on exposing the ways in which systems of oppression, inequality, and unearned advantage are racialized and inscribed by an ideology that places and supports "white" as the "normal," privileged, and most desirable racial identity (Clark and O'Donnell 1999; Delgado and Stefancic 1997; Fine et al. 1997; Frankenberg 1993; Giroux 1997; Grillo and Wildman 2000; Kincheloe and Steinberg 2000; Ladson-Billings and Tate 1995; Morrison 1992; Rodriguez 2000; Supriya 1999; Tatum 1999). Whiteness is, therefore, "a location of structural advantage, of race privilege . . . a standpoint and place from which white people look at ourselves, at others, and at society . . . (and) a set of cultural practices that are usually unmarked and unnamed" (Frankenberg 1993, p. 1). Critical whiteness guards against a systemic desire to be color-blind, which sets up a condition that "allows us to redress only extremely egregious racial harms, ones that everyone would notice and condemn" (Delgado and Stefancic 2001, p. 22). This colorblindness leaves untouched the ways in which racism is deeply engrained in how white people see and make sense of the world, allowing racism and marginalization to permeate everyday social practices.

Service-learning from a social justice perspective seeks to name those cultural and social practices that support systemic racialized inequality and privilege. Resistance to awareness of these privileging conditions has been theorized in the critical whiteness and white racial identity literature (Frankenberg 1993; Helms 1993; Kincheloe and Steinberg 2000; Tatum 1992; Thandeka 2001). For white students, service-learning often places them in an unfamiliar borderland that proves very threatening to the unearned advantages associated with being white. White students respond to this new knowledge with a range of emotions including guilt, anger, avoidance, and confusion (Arminio 2001; Helms 1993; Tatum 1997), which are often manifested as resistance. Using critical whiteness as a framework to think about service-learning and student resistance helps focus the analysis, in part, on the ways power and privilege work and are questioned or reinforced in educational practice.

In our experiences with undergraduate students engaged in service-learning, resistance, or oppositional behavior, is associated with a perceived threat to their positions of privilege and power and the subsequent need to maintain these positions. Never before confronted with their own unearned advantages (McIntosh 2001), we find white students resisting the critical and personal reflection necessary to produce new knowledge and awareness. Similarly, Kumashiro (2002) described this resistance to anti-oppressive education as a desire for repetition, the constant inscribing of what is known and taken-for-granted. Drawing from the work of Britzman, he explained "to learn in anti-oppressive ways, students need to do much more than learn that which affirms how they already understand themselves and what they already believe" (Kumashiro 2002, p. 70).

Our Praxis: Connecting Theory with the Practice of Service-Learning

As noted earlier, our service-learning courses are informed by the principles of critical pedagogy. As such, we work to make visible the ways in which knowledge is inscribed by a dominant ideology (e.g., whiteness) so as to raise consciousness about it and promote social, egalitarian, and transformative change (McLaren 2003a). Many of the tenets of critical pedagogy resonate with the goals of service-learning educators. McLaren (2003b) makes this connection explicit, suggesting that critical pedagogy is grounded in:

> . . . "walking the talk" and working in those very communities one purports to serve. A critical pedagogy for multicultural education should quicken the affective sensibilities of students as well as provide them with a language of social analysis, cultural critique, and social activism in the service of cutting the power and practice of capitalism at its joints. Opportunities must be made for students to work in communities where they can spend time with economically and ethnically diverse populations, as well as with gay and lesbian populations, in the context of community activism and participation in progressive social movements.   (pp. 170–171)

It is exactly these "opportunities" afforded by service-learning that bring students face-to-face with power and white privilege and create the conditions for student resistance.

READING RESISTANCE: "DON'T NOBODY BRING ME NO BAD NEWS"

The witch from the acclaimed Broadway show *The Wiz* captures our concern in offering student perspectives on their experiences with service-learning. On the surface, these quotations from written student reflections will *read* as a potentially negative commentary on their abilities to engage with the class. Further, these student "resisters" do not comprise the majority of the students who take the class. We also don't presume to know the basis for why resistance is performed by students in different ways; we do, however, see how resistance is performed via classroom and community contexts. We offer these profiles not as stable or discrete categories of resistance, but as illustrations of the complexities that emerge when engaging students in encounters with power, privilege, and the material conditions that produce the social inequalities in evidence at their service sites. What follows here are several thematic profiles of resistance and a critical developmental lens through which to examine them. In these stories of resistance, we find great authenticity and agency. We see authenticity in students presenting themselves exactly "how

I am" and agency in their struggle to make meaning of new experiences and knowledge. Through authenticity and agency, we seek to get *inside* the *underside* of service-learning.

PROFILES OF STUDENT RESISTANCE

We decided to explore the resistance demonstrated by students who seem to be academically ready for the course material, but who resist content and new knowledge for complex reasons. These are not the students who are challenged by the course material due to what we attribute to an issue of academic preparedness. We did not include these students in our analysis of resistance and instead focused on those white students for whom resistance seems to result from the complex intersections of the development of self-authorship and challenges to the previously unscrutinized privileging conditions exposed at their community service sites and through the readings for the course.

We purposely identified these student voices as "profiles" rather than types or categories because they are not clearly delineated, stable, or discreet units of analysis. Instead, the profiles, and the student perspectives they represent, are intended to capture some of the complex dynamics we experienced as students encountered the challenging environment of the service-learning classes and community settings. We also do not see these profiles from a hierarchical perspective suggesting that one response to resistance is "better" than another.

"The Good Volunteer"

"The Good Volunteers" are typically very good "helpers." They are usually very happy to help out at their service sites, although rarely discuss their experiences at the site as disrupting preconceived attitudes and beliefs about those with whom they come into contact. In fact, staff at the community service sites often comment that these students are their best volunteers. In class, they are generally quiet and acquiescent, in ways that might be interpreted by faculty as a silent, though begrudging, compliance or intellectual disengagement.

These "Good Volunteers" express their resistance through their written work and in class conversations that demonstrate an absence of critical thinking about the connections between complex social issues, power/privilege, and the very need for their community service work. When pushed to think in more depth, these students often react by asking the instructors and the service site leaders for more "contact" with the service recipients, as if being able to "see" a person living with HIV/AIDS will reveal some new insight. However, the "good volunteers" are not asking for a relationship with recipients, nor do they demonstrate any effort to get at the root of the social issues they are witnessing. Instead, they exhibit an attitude of entitlement in expressing that if they are going to help "those people" out, then they ought to be able to see first hand the results of their contributions.

"Good Volunteers" often expressed they should get more out of their experience at the service site than the individuals they are serving. In his article about the usefulness of service-learning, Butin (2003) writes that we need to be asking if our service-learning courses are ". . . a better comprehension of course content? . . . Or, . . . a voyeuristic exploitation of the 'cultural other' that masquerades as academically-sanctioned 'servant leadership?' " (p. 1675). We often receive requests to *see* the recipients of service from students who are working at a food pantry stocking shelves and loading boxes for needy families. One student stated, "The service site, [food pantry name], is really boring me. I just do manual labor. I don't get any interaction with people who come in . . . not that I'm ungrateful or anything." The question of gratitude for one's service comes up often from the "Good Volunteers." Implicit in the need to *see* the recipients is the expectation that students will receive thanks for their good work and individual effort.

In relation to class time and the service experience, some "Good Volunteers" indicated they enjoyed their service but saw no connection to the class concepts and theories. One student wrote, "I don't see why we have to do all the readings and papers. They don't seem to have anything to do with what I am doing at my service site." Another student expressed, "Reading is not one of my most favorite things unless it interests me. Some of our readings seem to be complaining and sob stories. I am sensitive to people with less fortunate conditions than myself, although I don't agree with overdoing it." These students are not actively resisting the material in the course; however, they do not see any relevance to their work in the community or to their own sense of responsibility. Most good volunteers appear to be passionate about performing charitable works. There is no resistance to the community service component of the class, but resistance emerges in the disconnect between the service and the objectives of the course.

"The Politely Frustrated" Volunteer

The "Politely Frustrated" students are similar to the "Good Volunteers" in that they do not wish to be disruptive and comply in order to get a good grade. The "Politely Frustrated" differ from the "Good Volunteers" in their emerging recognition of power structures at work at their service sites. The "Good Volunteer" may be well-intentioned but understands his/her work as helping out those "in need" or "victims" of bad luck. The "Politely Frustrated" students begin to see power and privilege, although they remain unapologetic about their own position of privilege (e.g., my family worked hard to get where they are).

The "Politely Frustrated" express their resistance through written work and course evaluations, rarely in public discourse, such as class discussions or in reflection groups. These students tell us in their final thoughts on the class that in their previous written work they were primarily telling us what they thought we wanted to hear, and holding firm to a view of education and service as apolitical. Their essays quietly express blaming service recipients for their own situations, no sense of individual responsibility for addressing complex social issues, and a lack of understanding between the connection between class readings/discussions and their work at the community organization. A student summed up this point of view in a course paper when he stated, "I do care about programs that help the less fortunate. I care about these as long as the people who benefit from them are trying to get work and get by and not just living by a check in the mail."

Another "Politely Frustrated" student wrote about feeling challenged when confronting issues of white privilege and systemic racism. She described feeling guilty and hurt after reading some of the course material, yet was unwilling to see this issue from any other perspective than her own. She believed that because she was not doing anything to personally oppress other people, systematic oppression did not exist. She wrote, "I don't think that just because I am white I have any more advantages than someone who is not white. I don't think that someone looked at my college application and said, 'oh, she's white, let her in.' . . . I received the same public education that everyone in the state of Ohio receives." This student and other politely frustrated resisters often feel attacked and hurt, and if they are not personally doing anything to perpetuate the system, they do not see their responsibility to confront the wrongs of the rest of the world.

While the "Good Volunteer" does not engage with the class, the "Politely Frustrated" students express that their time has been wasted during the class meetings and are increasingly frustrated by the service experience. One student, in the final course evaluation, offered some feedback, common among these students, about the course's reading materials. "The stories were awful. I got very little out of them and they had nothing to do with leadership, just about people's depressing lives . . . I liked my community service site but I feel like after this class I was supposed to feel so fulfilled that I made a difference, but I don't really think I did, nor did I really learn anything about myself."

Another student offered a common critique about the class content—the perceived focus on "diversity." Responding to the question about the most educational aspect of the course, this student wrote, "The service aspect. The classroom was not, in my opinion, educational. It just felt like most of the time it was 'OSU's diversity is great and all minorities and people that aren't middle class or upper class are like that way because they could not have helped it.' Which is so not true."

A male student who was a member of a selective, "leadership and service" scholars program took issue with the course and expressed strong resentment to course material that dealt with privilege and power. He identified himself as a "normal American" whose Christian values collided with some of the issues at his community service site, an AIDS service organization providing meals and nutritional support for people living with HIV/AIDS. His demeanor in the first hour was quiet and his body language, often sitting with his hands crossed at his chest and frequently, though quietly, sighing, communicated much discomfort.

In his final paper, he described his surprise in the times he let himself open up and learn while at the AIDS service organization. He could not, however, engage similarly during the class. In his final evaluation of the course he offered the following points of feedback. "I don't enjoy the preaching of a debatable agenda in the first hour. Perhaps teaching from a more balanced perspective would be better than 'isms are keeping us down.' " In response to what could be improved he offered, "More emphasis on community service. Less on ideologically driven readings and lessons."

"The Active Resister"

The "Active Resister" can be both hostile and disruptive throughout the entire service-learning experience. These students aggressively argue concepts presented in class in an open and confrontational way. They often try to garner support for their ideas (and critique of the class) from other students. These students also tend to be very problematic at the service site by making statements of blame pertaining to the service recipients and the situations in which they find themselves. The written work and class evaluations of these students make clear their opposition and resistance to the foundations of the course. These "Active Resisters" do not connect in any way with the course, not even at a superficial level.

A male student wrote in his evaluation, "I would not recommend this class to a friend. I would not allow my friend to have to listen to the things that are taught that are out of the spectrum of truth." Another student wrote, "You should never be forced to change your values. I would not recommend this class unless all you want to talk about is poor people." Finally, another student reflected on the readings about people struggling to live in the context of complex social issues and inequities and offered the following commentary about the insight he gleaned. "The people in the stories were worthless and inexperienced."

Another "Active Resister" was consistently angry about his involvement with the course. He was required to take the class for his "leadership and service" scholars program. We got off to a rocky start at the beginning of the quarter when he labeled a neighborhood adjacent to the University as "the ghetto." The students in the class had been brainstorming positive and negative attributes of a community and his "ghetto" comment was offered at the end of a discussion about the growing population of Somali people in the city. Given the oppressive racial and class implications of this label, the instructor immediately offered her perspective on what was damaging about the word "ghetto" and the ways in which such labels and stereotypes ran contrary to building capacity, which was the point of the exercise. White students and students of color seemed to take this opening as an opportunity to express their displeasure and offense at the use of his term as well. The student left the room feeling attacked and refused emails and personal approaches to check in with him. He did not, however, shrink from future classroom discussions and frequently interrupted students to share his point of view, naming it "the truth."

His final paper summed up his experience in the class and strong resistance to both the content and his work at the AIDS service organization:

> My main objection (to the class) came from the fact that the service seemed political in nature, more specifically, servant of a particular agenda endorsed by both the class and (the scholars program), which I resented.
>
> I wish this class would focus more on service and what really happens in the real world and not try to make all people look like victims when in reality it is mostly their own faults. When I am at [AIDS service organization] and I see a client, I know it was their fault they have AIDS, and if for some reason the person has AIDS by something not of their causing, they will be the very small percentage that is that way. If you choose a way to live, and the consequence causes you to be in a position where you need to rely on a non-profit organization to survive, then you must first realize it is your fault for that.

Another male student frequently shared in class, in his written work, and at his service site that this course was a waste of his time. He did not like working with children, though he chose to tutor at an after school program. Throughout the course and in follow-up conversations and evaluations after the course, he was consistent in expressing his frustration. He wrote:

> The best part of the course was that it was only one day a week. The most challenging was trying to answer questions that really didn't interest or pertain to me. It's hard to have opinions or feelings about something you aren't specifically passionate about. Like volunteering for a specific organization when (students in his leadership and service scholars program) in general just like helping out and volunteering, not because they want to strive to correct a social problem, just because they like helping out.

This student openly acknowledged that the focus of his service was his own self-gratification and no amount of reading, large group, or one-on-one discussion during the quarter would evoke any movement from this position. Finally, he felt that the best time to take this course was, "The first quarter of the freshman year, because they can just get it out of the way and forget about it."

READING STUDENT RESISTANCE: DISCUSSION AND IMPLICATIONS FOR PRACTICE

Where does resistance come from? What are the dynamics at work that foster the expression of resistance in the service-learning context? How might we *read* resistance so as to best understand the basis for its presence in service-learning and to transform resistance from a somewhat aggravating and potentially disruptive force to a site of critical thinking, creativity, and possibility? These are questions we ask ourselves when resistance appears in our service-learning classes (and it always does).

What makes *reading* student resistance in the context of service-learning so vexing is that it is not always easily discernable or translatable into practice.

Integrating the theoretical frameworks of self-authorship and critical whiteness provides one way to understand resistance because it focuses on *what* students are resisting (e.g., their own positions of power and privilege), *why* they are resisting (e.g., developmental readiness), and the *context* that sparks awareness of resistance (e.g., critical service-learning). Commenting on the cognitive restructuring necessary to promote racial/ethnic identity, Torres and Baxter Magolda (2004) aptly capture the importance of integrating theories of self-authorship with critical whiteness in understanding student resistance in service-learning:

> Although the *content* of reconstructing White privilege among majority students is very different than the *content* of reconstructing oppressive social images for ethnic and racial minorities, the *process* for both groups is dependent upon the interrelationship of cognitive and identity development. (p. 334)

Our *reading* of student resistance suggests that you cannot pull these constructs apart. To understand student resistance as only developmental obscures the powerful hold of privilege on student identities. To only focus on a critical whiteness approach fails to explain why some students grapple more readily with white privilege than others and their own complicity in the perpetuation of the structures of inequality they witness at their community service sites. Our findings suggest that student resisters in our courses do so in part because of deeply ingrained privileges, lack of exposure to those different from themselves, and absence of the developmental complexity required to recognize privilege in the first place.

No one template (despite the presence of many statements of "principles of best practice" in the service-learning literature) assures that all students will engage fully with the complexities of service-learning or that all service-learning educators will design experiences that "work." Further, much of what gets constructed as service-learning fits uncomfortably within a social justice approach and looks more like charitable work. What we offer here are a few implications for practice that grow directly out of our working through the real demonstrations of student resistance in our service-learning classes. These implications are informed by a critical developmental framework that integrates theories of self-authorship and whiteness.

Turning a Critical Developmental Lens on
Student Resistance

The framework of self-authorship reminds us that students come to our service-learning classes from different developmental places that influence what they see, know, and understand. It also suggests that many traditional aged college students are still relying on external sources that *author* them (Baxter Magolda 2000b). We see in our profiles the result of these dynamics as students resist new knowledge and experiences that counter what they are accustomed to. We also see that the experience of cognitive dissonance, prompted by their service-learning encounters, created some impetus to think differently, and more complexly, about their own identities in relation to what they were learning.

Because our service-learning course content is explicitly focused on complex social issues and the dynamics involved in their perpetuation, students' resistance can be understood as avoidance of any new knowledge that disrupts these external formulas and current ways of negotiating their place in the world. Some of our students found that engaging with service-learning becomes messy and tense, so a decision that class material is "irrelevant" creates a passage through which to avoid the conflict in the first place. "Irrelevance" can be interpreted as unwillingness to engage with questions of power and privilege or inability to understand the questions asked of them. This is both an identity and a cognitive issue. For example, the "Good Volunteer" does not seem to have a framework through which to analyze concepts of the course and therefore, only interacts with these issues in a superficial way and does not experience any challenges to the taken-for-granted (e.g., whiteness) assumptions about race. External formulas are currently working for them and they are not experiencing any cognitive dissonance.

To promote self-authorship through service-learning, a holistic approach to learning and development must be adopted. A holistic approach is one that incorporates "opportunities for complex cognitive, intrapersonal, and interpersonal development" (Baxter Magolda 2000b, p. 94). In particular, Baxter Magolda suggests: viewing students as capable participants in the journey to self-authorship, providing directions and practice in acquiring internal authority, establishing a community of learners among peers, and supporting the struggle inherent in exchanging older, simpler perspectives for newer, more complex ones. Further, we suggest that whiteness provides a conceptual framework for thinking about student resistance and self-authorship provides the pathway for analyzing resistance from a developmental perspective. We place our own reflections on implications within the context of these four themes integrated with a perspective from concepts from critical whiteness.

Viewing Students as Capable Participants in the Journey

The implication in this theme is the importance of engaging students where they currently are in their own development, rather than as Kegan (1994) exhorts, "We cannot simply stand on our favored side of the bridge and worry or fume about the many who have not yet passed over" (p. 62). The importance of respect and support for students' current realities cannot be underestimated in this work. This includes recognizing that many white students have never before been encouraged to examine their racial identities or the accompanying privileges. Because many of our students are still relying on external formulas, establishing an interpersonal connection with them is important as they construct meaning of their experiences (Ignelzi 2000) and learn to trust the experience they are having. Further, given the developmental readiness of many students and the role of external forces in self-authoring, it is important for service-learning educators to remind themselves of the power and responsibility that comes with serving as an authoring figure in students' lives.

Assessing early on the developmental readiness among students in service-learning classes may also prove useful in transforming resistance into an educational opportunity and promoting self-authorship. Incorporating into early assignments an opportunity for students to describe the nature of their previous community service involvement, views on service, and their motivations for taking the class may provide some clues. We incorporate into our classes an essay assignment that asks students to situate themselves in terms of their social identities and the intersection of these identities with community service. This provides insight into student self-knowledge regarding race, class, gender, sexual identities. Providing a mechanism for student choice in service site selection may also alleviate some of the risk of doing harm in the community, as well as provide instructors some information about how willing students are to engage with challenging community settings. Because students are expected to engage in the community service component of the class early in the quarter, they develop ownership for their own learning experience.

We suspect that providing students with multiple ways to engage with the course and express themselves helps educators understand where students are in their own development and then how best to promote new understanding and knowledge. For example, if students are resisting discussion about privilege in regard to race, the instructor might consider another social identity (e.g., gender, social class, religion) with which they might identify. Ongoing instructor feedback to student written reflections is another strategy for affirming students' developmental realities as they move through the borderlands of service-learning.

Providing Directions and Practice Acquiring Internal Authority

If service-learning educators strive to promote self-authorship and move toward anti-oppressive change, these efforts must be intentional, clearly explained to students, and reflected in all aspects of the course. Attention to the nature of student involvement in the community, class readings and discussion, and the connection between the two must be ongoing and deliberative. As summarized by Baxter Magolda (2000b) self-authorship is promoted through use of ill-structured problems (community service), exercises in perspective taking, dialogue on complex topics, personalizing learning through reflection, inviting students to say what they really think, and take risks in the classroom.

While the community service experience may serve as a catalyst to students' resistance, a space in the classroom must be made to enable students to voice their thinking. Because we have learned to expect resistance, the challenge becomes transforming resistance into an opportunity for dialogue and learning. One of the strategies to promote such learning and self-authorship is to facilitate discussions that do not silence the resisters while also not intimidating others in the class. Active resisters need to know that their voices are heard too, but sometimes follow-up with these more vocal resisters can't happen in front of the entire class, but instead outside of the classroom environment that may be less threatening.

The nature of the community service experience also is integral to promoting self-authorship. Without tapping into students' meaning making through active reflection, students are left to come up with their own analysis that may serve to reinforce inaccurate information and stereotypes. In addition, students may make comments at the community service site that are harmful and hurtful; and as one of our community partners stated, ". . . they just blurt out their opinions without thinking about who might be in the room with them . . . I really don't want to hear about it." Direct service and the opportunity to develop relationships with individuals at the community site are integral to shifts in thinking and awareness of power and privilege. The opportunity to personalize complex social issues and to see the real effects of social and public policies on the life situations of certain individuals does more to disrupt taken-for-granted assumptions than anything else we have designed into our service-learning courses. However, there is also a risk in this approach as some students may not be developmentally ready for what they may encounter at certain sites and retreat—or resist—new learning. As service-learning educators, it is important to negotiate the balance in asking community partners to potentially serve as a place where students make mistakes with the real contributions students can make in advancing the goals and activities of the community organization.

Establishing Communities of Learners Among Peers

For students relying on external formulas to make meaning, the support of peers is important (Ignelzi 2000). Peers provide both affirmation and challenges to move toward more internally defined definitions. At least half of each of our class meetings included small group reflection where students are grouped according to their community service sites. In these groups, students find a space to both speak up and defend their points of view as well as challenges to their ways of thinking by their peers. They often find themselves in the presence of multiple perspectives on issues that promote critical thinking and greater self awareness. We also have found that students are more likely to take risks if they see their peers engaging in new ideas and different experiences, rather than if the instructor suggested such activities. However, facilitators of reflection and class discussions must be attuned to the amount of "air time" some students take. Given the nature of the topics discussed in classes, some students may perceive this time as appropriate to air previously held opinions about issues rather than draw from what they have read or witnessed at their community service sites. Discussions and reflections must be designed so that all students may be heard and opposing viewpoints discussed in a space that is perceived as safe.

During the course of the quarter, students in our classes also participate in group projects structured around the issues at their community service site. This serves the dual purpose of establishing a community of learners working at a common community site and encouraging the development of multiple perspectives around complex social issues. For example, each small group must research and present, to the entire class, the social issues prevalent at their community service organization. The results of this research usually help students see these issues more complexly and begin to challenge more simplistic notions about why individuals are accessing the services of a particular organization.

Supporting the Struggle Inherent in the Journey

The concept of the "consciousness bridge" is useful to understanding this theme. The bridge foundation must support the journey across as well as invite students to make the journey in the first place. Students will quickly become "in over their heads," and resist, if course designs and expectations require capacities and skill in unfamiliar areas. In order to anchor the bridge on both ends *and* pay attention to the crossing, the structure of the design of service-learning must be "*meaningful* to those who will not yet understand that curriculum and *facilitative* of a transformation of mind so that they will come to understand that curriculum" (Kegan 1994, p. 62). Support for the journey is imperative. Inviting student reflections and providing feedback is one way to provide the support necessary to encourage new ways of thinking.

Further, when integrating self-authorship with critical whiteness, the "how I am now" is less a fixed sense of self, but instead one negotiated in relation to the contextual influences of the service-learning experience and the nature of the bridge supporting the crossing into the borderland. Student resistance in the context of service-learning cannot be understood without attention to both the foundation and the crossing over of the bridge. In this focus on both the structure and the crossing, we find the integral relationship between the development toward self-authorship and new awareness of power and privilege.

When white students first begin to acknowledge and understand some of the consequences associated with power and privilege, a full range of emotions can emerge (e.g., guilt, anger, sadness, hesitation). This is very tough work that often puts students into conflict with their families and peers. Instructors sharing their own journeys (and mistakes) help students appreciate that this is life-long work and cannot be easily resolved. Faculty anticipation of the emotional responses to such learning and development is important as is awareness of other resources around the campus and community that support student understanding of whiteness.

CONCLUSION: GETTING INSIDE
THE UNDERSIDE

Our experiences teaching service-learning courses suggest that the underside of service-learning is always present. Sometimes it lurks just beneath the surface and emerges tentatively and sporadically. Other times, it is very visible and present in how students navigate all aspects of the service-learning course. By bringing together a critical whiteness approach with a developmental one, educators have an opportunity to get inside the underside of service-learning and understand student resistance in new ways. Such an integrative framework acknowledges the developmental realities students bring to their service-learning experiences, the privileging conditions that situate students in service environments, the structures of inequality that produce the need for community service organizations, and the potential of service-learning as a critical pedagogy that opens up the possibility for anti-oppressive change. As bell hooks (2003) reminds us, getting inside the underside is ultimately a "pedagogy of hope," stretching the boundaries of educational practice to include "the life-enhancing vibrancy of diverse communities of resistance" (p. xvi).

REFERENCES

Arminio, J. (2001). Exploring the nature of race-related guilt. *Journal of Multicultural Counseling and Development*, 29: 239–252.

Baxter Magolda, M. B. (1999). *Creating contexts for learning and self-authorship: Constructive-developmental pedagogy.* Nashville, TN: Vanderbilt University Press.

Baxter Magolda, M. B. (2000a). Interpersonal maturity: Integrating agency and communion. *Journal of College Student Development*, 41: 141–156.

Baxter Magolda, M. B. (2000b). Teaching to promote holistic learning and development. In M. B. Baxter Magolda (Ed.), *Teaching to promote intellectual and personal maturity: Incorporating students' worldviews and identities into the learning process* (pp. 88–98). New Directions for Teaching and Learning, no. 82. San Francisco: Jossey-Bass.

Butin, D. W. (2003). Of what use is it? Multiple conceptualizations of service-learning in education. *Teachers College Record*, 105(9): 1674–1692.

Butin, D. W. (2005). Identity (re)construction and student resistance. In D. W. Butin (Ed.), *Teaching social foundations of education: Contexts, theories, and issues* (pp. 109–126). Mahwah, NJ: Lawrence Erlbaum Associates.

Clark, C. & O'Donnell, J. (1999). Rearticulating a racial identity: Creating oppositional spaces to fight for equality and social justice. In C. Clark and J. O'Donnell (Eds.), *Becoming and unbecoming white: Owning and disowning a racial identity* (pp. 1–9). Westport, CT: Bergin & Garvey.

Delgado, R. & Stefancic, J. (1997). Introduction. In R. Delgado & J. Stefancic (Eds.), *Critical white studies: Looking behind the mirror* (pp. xvii–xviii). Philadelphia, PA: Temple University Press.

Delgado, R. & Stefancic, J. (2001). *Critical race theory: An introduction.* New York: New York University Press.

Ellsworth, E. (1994). Why doesn't this feel empowering? Working through the repressive myths of critical pedagogy. In L. Stone (Ed.), *The education feminism reader* (pp. 300–327). NY: Routledge.

Evans, N., Forney, D., & Guido-DiBrito (1998). *Student development in college.* San Francisco: Jossey-Bass.

Eyler, J. & Giles, D. (1999). *Where's the learning in service-learning?* San Francisco: Jossey-Bass.

Fine, M., Powell, L. C., Weis, L., & Wong, L. M. (1997). Preface. In M. Fine, L. C. Powell, L. Weis, & L. M. Wong (Eds.), *Off white: Readings on race, power, and society* (pp. vii–xii). New York, NY: Routledge.

Frankenberg, R. (1993). *White women, race matters: The social construction of whiteness.* Minneapolis, MN: University of Minnesota Press.

Giroux, H. (1992). *Border crossing: Cultural workers and the politics of education.* NY: Routledge.

Giroux, H. (1997). Rewriting the discourse of racial identity: Towards a pedagogy and politics of whiteness. *Harvard Educational Review,* 67(2): 285–319.

Giroux, H. (2003). Critical theory and educational practice. In A. Darder, M. Baltodano, & R. D. Torres (Eds.), *The critical pedagogy reader* (pp. 27–56). NY: RoutledgeFalmer.

Grillo, T. & Wildman, S. M. (2000). Obscuring the importance of race: The implications of making comparisons between racism and sexism (or Other –Isms). In R. Delgado & J. Stefancic (Eds.), *Critical race theory: The cutting edge* (pp. 648–656). Philadelphia, PA: Temple University Press.

Hayes, E. & Cuban, S. (1997). Border pedagogy: A critical framework for service learning. *Michigan Journal of Community Service-Learning,* 4: 72–80.

Helms, J. E. (1993). Toward a model of white racial identity development. In J. E. Helms (Ed.), *Black and white racial identity: Theory, research, and practice* (pp. 49–66). New York: Greenwood Press.

hooks, b. (2003). *Teaching community: A Pedagogy of hope.* New York: Routledge.

Ignelzi, M. (2000). Meaning-making in the learning and teaching process. In M. B. Baxter Magolda (Ed.), *Teaching to promote intellectual and personal maturity: Incorporating students' worldviews and identities into the learning process* (pp. 5–14). New Directions for Teaching and Learning, no. 82. San Francisco: Jossey-Bass.

Jones, S. R. (2002, September/October). The "underside" of service-learning. *About Campus,* 7(4): 10–15.

Jones, S. R. & Abes, E. S. (2003). Developing student understanding of HIV/AIDS through Community Service-Learning: A case study analysis. *Journal of College Student Development,* 44(4): 470–488.

Jones, S. R. & Abes, E. S. (2004). Enduring influences of service-learning on college students' identity development. *Journal of College Student Development,* 45(2): 149–166.

Jones, S.R. & Hill, K. (2001). Crossing High Street: Understanding diversity through community service-learning. *Journal of College Student Development,* 42: 204–216.

Kahne, J. & Westheimer, J. (1996). In the service of what? The politics of service-learning. *Phi Delta Kappan,* 77: 592–598.

Kegan, R. (1994). *In over our heads: The mental demands of modern life.* Cambridge, MA: Harvard University Press.

Kincheloe, J. L. & Steinberg, S. R. (2000). Constructing a pedagogy of whiteness for angry white students. In N. M. Rodriguez & L. E. Villaverde (Eds.), *Dismantling white privilege: Pedagogy, politics, and whiteness* (pp. 178–198). New York: Peter Lang.

Kumashiro, K. K. (2002). Against repetition: Addressing resistance to anti-oppressive change in the practices of learning, teaching, supervising, and researching. *Harvard Educational Review*, 72: 67–92.

Kumashiro, K. K. (2004). Uncertain beginnings: Learning to teach paradoxically. *Theory into Practice*, 43: 111–115.

Ladson-Billings, G. & Tate, W. F. (1995). Toward a critical race theory of education. *Teachers College Record*, 97: 47–68.

McIntosh, P. (2001). White privilege and male privilege: A personal account of coming to see correspondences through work in women's studies (1988). In M. L. Anderson & P. H. Collins (Eds.), *Race, class, and gender: An anthology* (4th edition) (pp. 95–105). California: Wadsworth/Thomson Learning.

McLaren, P. (2003a). Critical pedagogy: A look at the major concepts. In A. Darder, M. Baltodano, & R. D. Torres (Eds.), *The critical pedagogy reader* (pp. 69–96). New York, NY: RoutledgeFalmer.

McLaren, P. (2003b). Revolutionary pedagogy in post-revolutionary times. In A. Darder, M. Baltodano, & R. D. Torres (Eds.), *The critical pedagogy reader* (pp. 151–184). New York, NY: RoutledgeFalmer.

Morrison, T. (1992). *Playing in the dark: Whiteness and the literary imagination.* New York: Vintage.

Morton, K. (1995). The irony of service: Charity, project and social change in service-learning. *Michigan Journal of Community Service-Learning*, 22: 19–32.

Nieto, S. (2000). Introduction. In C. O'Grady (Ed.), *Integrating service learning and multicultural education in colleges and universities* (pp. i–xiii). Mahwah, NJ: Lawrence Erlbaum.

O'Grady, C. R. (2000). Integrating service learning and multicultural education: An overview. In C. R. O'Grady (Ed.), *Integrating service learning and multicultural education in colleges and universities* (pp. 1–19). Mahwah, NJ: Erlbaum.

Perry, W. G. Jr. (1978). Sharing in the costs of growth. In C. A. Parker (Ed.), *Encouraging development in college students* (pp. 267–273). Minneapolis: University of Minnesota Press.

Rhoads, R. A. (1997). *Community service and higher learning: Explorations of the caring self.* Albany: State University of New York Press.

Rodriguez, N. M. (2000). Projects of whiteness in critical pedagogy. In N. M. Rodriguez & L. E. Villaverde (Eds.), *Dismantling white privilege: Pedagogy, politics, and whiteness* (pp. 1–24). New York: Peter Lang.

Rosenberger, C. (2000). Beyond empathy: Developing critical consciousness through service-learning. In C. R. O'Grady (Ed.), *Integrating service learning and multicultural education in colleges and universities* (pp. 22–44). Mahwah, NJ: Erlbaum.

Supriya, K. E. (1999). White difference: Cultural constructions of white identity. In T. K. Nakayama & J. N. Martin (Eds.), *Whiteness: The communication of social identity* (pp. 129–148). Thousand Oaks, CA: Sage Publications.

Tatum, B. D. (1992). Talking about race, learning about racism: The application of racial identity development theory in the classroom. *Harvard Educational Review*, 62(1): 1–24.

Tatum, B. D. (1997). *Why are all the Black kids sitting together in the cafeteria?* New York: Basic Books.

Tatum, B. D. (1999). Lighting candles in the dark: One black woman's response to white antiracist narratives. In C. Clark & J. O'Donnell (Eds.), *Becoming and unbecoming white: Owning and disowning a racial identity* (pp. 56–63). Westport, CT: Bergin & Garvey.

Thandeka (2001). *Learning to be white: Money, race, and God in America.* New York, NY: Continuum.

Torres, V. & Baxter Magolda, M. B. (2004). Reconstructing Latino identity: The influence of cognitive development on the ethnic identity process of Latino students. *Journal of College Student Development,* 45: 333–347.

Weiler, K. (1988). *Women teaching for change: Gender, class & power.* South Hadley, MA: Bergin & Garvey.

2

"Whose School is it Anyway?" Student Voices in an Urban Classroom

Raji Swaminathan

Introduction

How do we engage youth in school and in the community? To what extent are student voices taken into account in structuring educational experiences? What can we learn when students' perspectives are taken seriously in schools? Engaging students in school and the community and encouraging youth to believe in the "power of their ideas" (Meier 1995) are twin concerns of educators. Service-learning has been widely embraced as a way to engage students in school and community while critical pedagogy has been advocated as a means of encouraging student voices in the classroom. In this chapter, I describe a classroom at a high school where students reflected critically on their service-learning experiences as they questioned, analyzed, and critiqued the assumptions of what it means to do service. The resulting data show that for youth to be empowered through service-learning, it is crucial that their views be taken into account in structuring meaningful experiences. Second, the data reveal that creating spaces where students can reflect on their experiences within a critical multicultural framework provides a dynamic space of hope and possibility in schools.

Silencing or Honoring Student Voices

The terms critical multiculturalism and critical pedagogy are not identical, nevertheless, I use them interchangeably after Geneva Gay (1995) who argues that on common issues the two are "mirror images." Gay (1995) points out that on issues of educational access, equity, and excellence in a pluralistic society, the two movements share similar concerns and perspectives. Practitioners of both critical multiculturalism and critical pedagogy attempt to open up spaces for students to examine and critique relationships of dominance and subjugation (Sleeter and McLaren 1995) as well as reflect on the social, political, and economic forces that have helped to shape their lives. In this chapter, I describe a classroom where such a space was created for high school students to share what Fine and Weis (2003) refer to as "extraordinary conversations" as they reflected on and interrogated their community service-learning experiences.

Scholars have pointed out that far from honoring student voices, mainstream schooling tends to silence them (Fine 1991; Greene 1986), controlling formally and informally "who can speak, what can and cannot be spoken and whose discourse must be controlled" (Fine 1991, p. 33). According to Fine and Weis (2003), schools tend to silence students not only by evading some topics but also by systematically expelling voices of critique and rendering them deviant when such discourse is dissenting or challenging of official school knowledge. Scholars (Fine and Weis 2003; Giroux 1993) have argued that such silencing serves to reproduce and maintain social class and race hierarchies in schools that mirror the economic, civic, and social divisions of society. While examples of such silencing are not difficult to find, Fine and Weis (2003) also paint sketches of spaces in schools where "extraordinary conversations" take place, where teachers and students explore the "knowledge and wisdom that urban youth both bring into and can construct within public schools, if only we would listen" (p. 7). Like Fine and Weis, other advocates of student voice have pointed out that student narratives of their experiences provide a rich source for engaging and understanding youth (Quiroz 2001). Student participation that engenders Fine and Weis' "extraordinary conversations" is not the result of attempts to garner periodic comments or feedback from students or to use participation as a cover for student compliance. They are the result of empowered voices that incorporate relationships of reciprocity (Ruiz 1997). Richard Ruiz (1997) explains that for voice to be empowering it has to be heard, not simply spoken. Ira Shor (1996) makes a similar point when speaking of empowered classrooms, where students occupy the "enabling center of their educations, not the disabling margins" (p. 200). However, empowerment within critical pedagogy does not restrict itself to celebrating the expression of voice but includes interrogating the experiences and interpretations of youth. Indeed, critical pedagogy assumes that student voices are framed within a society that privileges the stories, meanings, and values of some over others.

However, when student voices are taken seriously, the results are encouraging. Students are astute observers of their educational environment (Rudduck et al. 1996; Soohoo 1993), and their views on teaching and learning are found to be very similar to views of contemporary educational theorists (Phelan et al. 1992). Further, Nieto (1994a) found that students could not only instruct curriculum but more important be catalysts for change by challenging teachers to confront their unintentional biases. Nieto's point ties in with the project of critical multiculturalism that seeks to move away from the culture of silence that dominates mainstream classrooms to create room for critical dialogue where the analysis of power relations can be central to making "meaning" a term that under modernity is often seen as ahistorical, neutral, and separate from value and power. In the study I describe in this paper, a curriculum structured around community service-learning offered opportunities for students to reflect on their experiences and for staff to pay attention and respond to student voices.

Research Questions

My purpose was to investigate what takes place when spaces are opened up for students and teachers to have a dialogue on issues that go beyond the explicit curriculum of the school to topic areas that are less defined but more immediately relevant to the lives of students. In this case, the study examined how students interpreted and analyzed their community service-learning experiences.

The questions that focused my observations and directed the study of this classroom were:

(1) How do students understand, critique, and analyze their community service-learning experiences?
(2) What can educators learn from student views of their educational experiences?
(3) What changes take place when educators take student perspectives seriously?
(4) How can educators design appropriate service-learning experiences for a diverse student body?

Methodology

This study was part of a larger ongoing study on how innovative schools serve students at risk of failure in traditional settings. The data reported in this chapter were gathered over one semester at an urban high school in a seminar called "What's in a Name." Data were gathered through participant observation. I observed 22 hours of class and took notes of all discussions that comprised the bulk of the data. Data included student reflective journals and interviews with the teacher as well as informal chats with the students. My continued long-term presence at the school together with interviews with staff and alumni augmented the data gathered by helping me understand the larger context within which the classroom was framed. As each class session was completed, I transcribed, coded, and carried data into the subsequent classes and freely shared the notes with the students. The analysis proceeded along item, pattern, and structural levels (LeCompte 1999).

The site chosen for this study was the Community School,[1] situated in a busy downtown urban area, for its "innovative" or alternative status and mission to educate using community service-learning. Additionally, the Community School has a record of high attendance and graduation rates.

While the school district data show an average truancy rate of 50 percent, the Community School has an attendance rate of 90 percent. The data on graduation is similar compared with a 50 percent dropout rate in the school district. The Community School draws its students from across the city and has the same demographic composition as the district.

MIDWEST CITY AND
THE COMMUNITY SCHOOL

The School Reform News (January 2002) reported that even for states that had high overall graduation rates, the rates for graduating minority students was poor. This is especially true for the large midwestern state where this study is situated. It has the second-best overall graduation rate (87 percent), but the worst graduation rate for African Americans (40 percent). Some cities in the state had a particularly bad record of graduating minority students. Midwest City has been a especially severe case where in 1995, 11 out of the city's 12 high schools graduated less than half of their ninth grade students. The city school district—referred to as Midwest City School District in this chapter—has an 80 percent student of color population, 61 percent African American, 21 percent Anglo-American, 14 percent Latino, 4 percent Asian, and 1 percent Native American. The district has a dropout rate of 50 percent (Barndt and McNally 2001).

Among several initiatives in the school district designed to meet and resolve the problem of attrition and reduce the failing rate in high schools, one initiative called for creating alternatives to the existing systems of learning through the Innovative Schools program. In the year 1995, two veteran teachers—Mark and Larry, with 20 years of classroom teaching experience behind them, answered the Midwest School District call for proposals for small schools that would outline new curricular approaches to learning. Mark, a Social Studies teacher, and Larry, a Math teacher, visited a variety of alternative schools and finally decided to model their school after a program in New York City called the City-As-School project. Running successfully for over fifteen years, the City-As-School model became the inspiration for founding the Community School.

CITY-AS-SCHOOL MODEL AND
THE COMMUNITY SCHOOL

The City-As-School was founded in New York City in 1972 with a staff of 4 and 15 students. Over the years, the school has grown to 1200 students with a faculty of 104 and operates out of 3 sites. The goal of the City-As-School was to link students with learning experiences throughout the community that included internships, apprenticeships, and mentoring by professionals in businesses and service organizations. Students who completed the requirements for each learning experience were awarded credit toward high school graduation. The area of "service" as defined by the City-As-School project encompassed internship experiences at public service institutions, a focus on learning specific skills, working with community centers and community groups to research problems and participate in community service projects.

Larry and Mark first came across the City-As-School model at a conference presentation on successful models for youth placed at risk. Impressed with what they learned, they visited the school and returned energized. Their visit sowed the seeds of a collaboration and partnership that continues to endure. Every year, as part of a staff development initiative, teachers from the Community School travel to New York and spend a few days exchanging ideas, brainstorming, and problem solving with the staff at the City-As-School project.

In setting up the Community School, the founding teachers followed the model of the City-As-School by developing internships across the city at several sites—professional offices, schools, hospitals, trade centers, and community agencies that would collaborate on structuring learning experiences for students with the idea that experiential learning would serve to link classroom texts and abstract concepts with real life, make learning more enjoyable, and allow students to participate in making decisions regarding their own education. A second idea was that community service became a central theme of the structured learning experiences of the students. These two ideas together formed the backbone of the curriculum.

In the late spring and early summer weeks of 1995, Mark and Larry were joined by three teachers interested in putting their energies into "a different way of teaching and learning." The five teachers together founded the Community School. All through late spring and summer they held regular meetings, drew up alternative assessments, designed experiential learning modules, recruited students from across the city, and opened that first year with a total of 86 students and 5 teachers and 1 additional staff member. The school comprised 11th and 12th grades only. They proposed that it be a teacher-run school with one teacher-in-charge position. The teacher-in-charge would be part administrator but the entire group would function using a process of democratic decision-making.

MAKING COMMUNITY CONNECTIONS

The most exciting and labor-intensive part of setting up the school was making connections with and setting up sites where students could intern for either nine weeks or a semester at a time. The teachers called on their familiarity with the city and its community groups to find support for the new curriculum that they proposed. The process of setting up a site began with a teacher having an idea, bringing it to a faculty meeting, and unless objections were raised, went on to build a collaborative relationship and obtain assent from key players at the site. For example, one of the teachers, Mary, set up a meeting with a local dance and theater group. She took school brochures, photographs, examples of student work at other sites, and made a presentation about the school. Once the group at the site was interested, discussions followed regarding the roles of the teacher, the site, the student, and the supervisor who would work with the student placed at the site. Examples of student work for academic credit included research, observation, and report-writing. However, while it was important that these aspects be included, the types of assessments and the areas in which academic credit was received were to be determined in collaboration with the site supervisor and depended on the focus of the internship. If the student at the theater site was involved primarily with lighting, it was possible to work out an assignment that involved special effects and its influence on modern theater, or an assignment that detailed the technical aspects of lighting and sound. The teachers created a standard tool and format for the assessment of student work. This was the LEAP or the Learning Experience Activity Packet that demanded journaling and research in addition to a fixed number of hours of experience at the site. The school developed over a hundred sites for community service-learning projects.

The school schedule accommodated community service-learning as the central piece of the curriculum. Teachers taught in-house classes on two days and visited students' sites on two. On Fridays, all students and staff were present in the building. Fridays were used for a series of seminar classes. Seminars were frequently student initiated on topics of their choosing. Any given semester saw eight seminars and the semester I gathered data, they included one on community problems, one on science fiction film, a book club, a journalism class, and one on Greek myths led by two students. The seminar I attended was called "What's in a name." It was facilitated by the math teacher, also one of the core founders at the school. The seminars were followed by an advisory period where students and teachers had an opportunity to discuss issues, chat and work out where things stood with regard to each student's academic and site progress. Additionally, on Fridays one could also see some members of the alumni either talking with the students or making brief, formal presentations giving students tips on getting to college or acquiring a job.

THE COMMUNITY SCHOOL TODAY

Nine years after it was founded, in the year 2004, the Community School had grown to hold 127 students and 8 teachers. The school continues to serve the 11th and 12th grades and students' ages ranged from 16 to 19 with the majority being 17 and 18 years old. The student population mirrored the district breakdown in terms of race. The teacher composition changed with retirements and the four female and four male teachers comprised three African Americans, one Latino, and four Anglo-Americans among them. Of the original five founding teachers, three were still on the staff, and Mark was the teacher-in-charge. Admissions criteria included an interview with the student and an additional interview with either a parent or guardian, and roughly 10 high school credits.

BACKGROUND TO THE SEMINAR

The seminar "What's in a name" was offered by the math teacher, Larry, in response to direct requests from two groups of students following murmurs around the school regarding how some students felt positioned in relation to community service-learning activities. Jokes, laughter, and at times a hint of indignation accompanied casual student comments on how their families or friends at other schools interpreted community service-learning. One student mentioned half jokingly—"some people think I am a juvie" (juvenile delinquent). Although such comments were occasional, they had begun to crop up more frequently and the teachers at the school were anxious to find out and resolve what was clearly troubling students. A group of students began discussing informally among themselves and within their advisory groups, the possibility of changing the name of the school. Advisory groups were set up so that a small group of students and a teacher met every Friday for 45 minutes to discuss the week gone by and the week to follow. This was when students raised issues that concerned them and when they could chat casually with the teacher and their peers about life in general, the neighborhood, exams, family, and friends. One advisory group in particular saw the issue brought up frequently. According to the students in that group, the name of the school was the crux of the problem. Attending the Community School and doing community service sent the message that they were delinquents who had exchanged jail time for community service. Students began to think that a different school name would make clearer to the community how the school was innovative rather than punitive. Two groups of students petitioned separately for a new seminar—one that would focus on changing the name of the school and a second where they could discuss the community service-learning projects with a view to making changes. Since both requests were primarily to do with questioning the assumptions behind community service-learning, Larry, the math teacher at the school, suggested incorporating them into a single seminar.

The seminar would interrogate the multiple meanings of community service-learning, examine student experiences, discuss the changes they wanted, and create an action plan to effect change. The seminar would provide a space where voices of critique could be raised without fear, and where students could experience democratic learning. Teachers were enthusiastic about the seminar and applauded Larry's choice of title, "What's in a Name?" The goal of getting student perspectives on school was in keeping with the school philosophy that stressed continuous self-improvement and reinvention. My presence at the staff discussion was serendipitous and I received permission to sit in on the seminar for that semester. Larry welcomed me into his classroom. Altogether 22 students signed up for the seminar, 20 of whom attended regularly. From the beginning, it was clear that the students had the freedom to determine the direction of the seminar. The students adopted a structure that appeared comfortable and familiar to them. Every class session, two students volunteered to take notes on the chalkboard and moderate discussions.

While every discussion in the seminar shed light on some aspect of student experiences of school life, two discussions are particularly significant for the questions examined in this chapter. In the rest of this chapter, I focus on these two discussions and summarize key contextual occurrences for ethnographic continuity. The first discussion was about how African American and Latino students had fundamentally different experiences from Anglo-American students in terms of how service was viewed and the second was a debate by students as to whether they should keep the results of a survey to change the name of the school.

COMMUNITY SERVICE-LEARNING: WHAT'S IN A NAME?

One of the most significant discussions that occurred in the seminar drew attention to the sharp contrast between the experiences of African American and Latino students on the one hand and Anglo-American students on the other on how community service-learning was viewed. Students brought up two major issues. The first had to do with how they felt positioned when they came in contact with the community. The second dealt with the ways in which community service-learning was structured and the extent to which student expectations were being met.

African American and Latino students raised issues of stereotyping, social status, and misidentification resulting from the ways in which people saw the school or read community service. The following excerpt illustrates their concerns.

> Tianna: When I told my cousin I was at the Community School and we do community service, he said—isn't that for juvenile delinquents who have to serve community service there as part of their court order?
> Ben: I just heard this the other day. No employer is going to want to give you a job. They're going to think you are a bad student who is out of the old school for doing something wrong.
> Larry (teacher): How many people in class have had this experience when you tell people you are doing community service as part of schoolwork?
> LaTora: We don't always say—we are doing community-service-learning. We usually just say—oh, I am doing this as part of my schoolwork or we just say we do community service at this school.
> Larry (teacher): Okay, so they hear you say you are doing community service. So what happens then? How many people in the class have had a negative response to the term?

About 70 percent of the class raised their hands, most of whom were African American, or Latino. Only two White students raised their hands and two Black students did not. Larry turned to one of the students who did not raise his hand and asked about his experiences.

> David (among the 30 % White students taking the course): I never experienced anything negative. I have heard things like—wow, that's cool. You're a credit to your age and stuff like that.

Lisa (White student): Yes, service-learning is everywhere now and it looks good on the resume. And if I say I am doing community service then people think—wow, she is really a good person or something like that. Nothing bad or negative.

John: Only if you are White. If you are a minority, they think you are doing it to cut down on jail time. People think I was involved with the police. Community service has a bad reputation among people if you are African-American.

Jesus: Or Latino.

Mira: I think it also depends on the type of community service we are doing. I was over at the homeless shelter and the people coming in to volunteer there first thought I was doing this because I was court ordered to do it. I was doing stuff like organizing the kitchen there and taking inventory and cleaning. The cleaning up especially makes people think either you are doing this because you have been ordered to do it or else why would you be there?

John: Which actually is something I want to ask about. Shouldn't all community service-learning sites teach us skills and academics as well? What credit should you get for cleaning up?

Mira (in a protesting voice): I got my academic credit for social studies because I did a project on the homeless in the city and calculated from data what percentage cycle through and return in the same year. Also I found out what skills do they need to learn in order to not be homeless. Cleaning was part of the community service and learning how the place works. I did interviews and also created a new record keeping system.

Larry (Teacher): Okay. And you are saying that the volunteers who came in there thought you were delinquent?

Mira: Yep. One woman actually asked me—what did you do honey? You seem so nice. And I was like—what do you mean? And she said—aren't you here because the judge told you ? Then my supervisor there—she overheard part of this, came over and explained to her and to all the volunteers about the school and what I was doing there and that I was a regular student not a "bad" student.

Octavia: Not a trash picker-upper.

Larry (teacher): So two points are being raised here—one is that some students from this class and I think from this school, especially students of color are being thought of as delinquents and the second is that the "type" of community service that you are engaged in may also lead to a similar perception.

There was general agreement in the class that this was the case. Between the above discussion and the next discussion excerpt that I present from my field notes, several events occurred. Larry steered the class discussion so that students could question the process of naming and being named. The students framed the issues of being labeled as a consequence of ignorance, prejudice, and media influence. They discussed how to create better awareness in the larger community, and initiated a plan of action to change the name of the school.

The seminar acquired an intensity and students appeared ready to take action. The tenor of the class changed from co-facilitation by the students and teacher to the students now taking charge of the seminar. Students decided that a democratic process needed to be followed. Next, they designed a survey with seven possible names for the school and a space for students to suggest a new name.

Before turning to the survey results that were the focus of a long debate among students, the class discussions thus far yielded two significant points. The first was that students desired greater input into structuring community service-learning projects. Second, there was a difference between what students of color wanted out of placements and what White students sought. Students of color stressed learning specific skills so that the emphasis was more on learning and less on service. They wanted community service to be folded into learning so that service did not become an end in itself. White students on the other hand wanted to continue with experiences where service was central and where they had the opportunity to interact with people from different cultures. For many White students, despite being at a predominantly minority school, it was only at experiential learning sites that they had the opportunity to work closely across racial divides. The class then turned to the results of the surveys on changing the name of the school that proved to be a ground for a different debate.

STUDENTS AS CHANGE AGENTS:
LESSONS IN DEMOCRACY

Students in the seminar awaited the results of the survey eagerly. Anthony, one of three students who had tabulated the results, drummed his fingers and promised that the hour was going to be "real loud." The results revealed that seniors preferred the existing name to any of the choices presented and were less invested in changing the name of the school than the juniors. Although disappointed, it was on hearing the winning name from the surveys that the class went into an uproar. The name chosen by most juniors was Midwest Independent High School much to the astonishment, disappointment, and irritation of the students in the seminar. Most students in the seminar favored the name: The Innovative School. The unexpected results led to their desires colliding with the process they had adopted. They wondered aloud if they could send out a second survey. Larry intervened to suggest that the process of democracy was not easy and challenged them to remain true to the process. The students rebelled—"*whose school is it anyway? And whose seminar is it?* Why should we do what everyone says? It is our education at stake here." Larry did not back down and instead continued to challenge them to consider whether in the process of becoming empowered they were silencing others in the school. Saying this, he left the room so that students could talk among themselves.[2]

The students' first response was to brush off Larry's words but 20 minutes later, after the students had debated whether or not to redistribute the survey, the discussion circled back to the issue of doing what was democratic. They agreed among themselves that staying the democratic route was the best and that perhaps the name change itself was less important than the process of including everyone in the dialogue.

Ricardo: Okay, everyone, what do we do here now?

Jessica: I think that the teacher is a small minority in this classroom and this is our seminar. We have the right to conduct another survey if we want to.

Octavia: Let's get a paper shredder and shred the first survey. There were lots of people who tried to influence others into voting for the name Independent High.

Carlos: So you are saying that the process of the first survey was not correct?

Octavia: Yes, that is what I am saying. It was not fair.

Dan: I think it was fine and it should stay the way it is. Are we saying we want the democratic process or not?

Octavia: That's because you thought of the Independent crap.

Maria: I think we should include everyone and maybe talk about this more in all Advisory groups. It could be that other Advisory groups didn't get the chance to talk about the pros and cons the way we did in our seminar.

Dan: We should have waited to give out the surveys. We first should have had an all school meeting and presented the pros and cons that came up in our class—and not just the name change but everything that came before it.

Octavia: Yes, some students didn't even know why they were voting. I think we need more talks and then an action plan.

Carlos: that is fine, but what do we do now? We don't want to do the wrong thing here because we wanted to do everything democratically. At this school, we all get to participate—we are able to have this seminar because of that— so is it right for us to do what we want instead of what we said we would do?

Octavia: I don't care—I want to shred the survey.

Maria: Think for a moment, I don't want to be like a bad government you know, say one thing and then hold elections and if I don't like the result, let's change it or hold another till I like the result.

Dan: And we all know how often that can happen.

Octavia: I am beginning to feel sympathetic for that now. Okay not seriously. But okay, I don't want to be like that either. I don't like it but I guess I have to agree to it. (In a more cheerful voice she added) I will graduate anyway before the Independent crap gets put in place.

Students talked among themselves after which they agreed that they did want to honor the process and be democratic even if they were unhappy with it.

The students called Larry into the room and reported to him on the actions they had decided to take.

Dan: Mr. M., we decided that you are right, but it is hard for some of us as we don't agree with the results. We wanted the school to be called the Innovative School. But we have decided to make a recommendation to the faculty and the student body seeing as this seminar is about to end and we would need a new seminar to carry out the work or need volunteers to take the school name change forward.

The seminar ended with students writing about what they learned and to what extent they thought it answered some of their questions and concerns. Most students agreed that the seminar had allowed them to discuss what they thought about community service-learning and how they felt positioned by the community while working at placements. They now thought they understood why people may create labels and most students felt confident that they could resist such labels. They also felt heard by the school faculty with regard to their ideas and concerns although as most students pointed out in their writing, the seminar raised difficult issues for them. Carlos's words are an example of what many students wrote—"We were tempted to not stick to the democratic process that we had adopted in the beginning. This was a valuable lesson in learning about how difficult it is to be responsible." The seminar concluded but left an impact on the school and teachers, raising questions on ways in which community service-learning needs to be examined, reflected upon, and restructured. In the next section, I analyze the ethnographic excerpts in this chapter around the research questions that framed this study.

DISCUSSION

Student Perceptions of Community Service-Learning

Critical researchers of service-learning have drawn attention to the power differential that exists among those "providing" service and those "receiving" it (see e.g., Purpel 1999; Wade 1997). They caution students engaged in service-learning to be aware of the power and privilege inherent in their position and of the danger of unwittingly reinforcing stereotypes in place of disrupting them. However, such research often casts the hierarchy of service-learning within a framework of privilege that favors service providers. This study indicates what some scholars like Varlotta (1997) have pointed out; that the self in relation to power in service-learning is more complex where "it is not the case that the student 'server' is permanently in the dominant position over the 'servee.'" This study found that students involved in service-learning might be positioned differently in unequal and multiple power relationships.

The first data excerpt in this chapter helps us understand how students perceived and analyzed their service-learning experiences. While scholars on service-learning have drawn attention to the positive effects of service-learning in building relations with the community (Melchior 2000), the insertion of race complicates what Davis and Harré (1990) refer to as "positioning" in service relationships. Davis and Harré (1990) use the term to refer to the ways in which individuals are constituted and reconstituted through social interactions and discursive practices. In human interaction, one may position another in a storyline that the other may or may not take up and may either choose to accept or resist.

The class discussion revealed that race played a significant role in the way students experienced being positioned in their community service-learning sites. While some scholars argue that race is an illusion (see e.g., Appiah 1992) and others that race takes on varied and multiple forms (Bartolome and Macedo 1997), there is nevertheless general agreement that race is critical to identities, both voluntary and involuntary (Dolby 2000). Involuntary identities or as Beverly Tatum (1997) refers to them, "targeted" identities are those that mark us as "other" within a social, economic, and political hierarchy of meaning that translates as "less than." This study found that students of color were marked as "other" even when they were engaged in the same acts of service as White students. Students in the seminar came up against a widespread perception regarding who does service and why that intersected with views of youth of color as "a generation of suspects" (Giroux 2003, p. 557). While students of color were misidentified as criminals and their participation in community service perceived as restitution,[3] community service on the part of White students was seen as positive and they were seen as a "credit" to their age. Taken together, both these perceptions serve to reproduce an environment of inequality that devalues students of color.

Students in the class recognized how their position in the social hierarchy was mediated by the intersection of race (Roschelle 1997). They also held that how one is categorized is at times dependent on the perceived status of the occupation or activity in which one is engaged. For youth of color who had been defined exclusively in negative terms within school, service-learning offered opportunities to create a different sense of self. However, as students' discussions revealed, the act of positioning in service-learning could lead to being negatively judged, and the judgment internalized, vitiating a person's sense of self-worth.

The trust between students and teachers at the Community School allowed students of color to address these issues and resist the ways in which they were being positioned. By doing so, students rejected internalizing negative stereotypes, sought the support of teachers, peers, and the institution in fighting covert forms of racism manifested in a societal perception of youth of color as criminals.

During the discussion, White students came to understand the role of race in service-learning activities. They were stunned to find that their peers were regarded as juvenile delinquents while, in contrast, they were seen as being responsible youth. The seminar presented an opportunity for them to question their own privileged status in society and find a ground where they could stand for social justice. If the seminar offered students the opportunity for cross-racial dialogue, it also called to question teachers' taken-for-granted assumptions around equity and fairness.

LEARNING FROM STUDENT VIEWS

The seminar presented a dynamic space where students took a step into the realm of community organizing by creating action plans, debating how problems are resolved and making room for all voices to be included. From an analysis of the data, we begin to see how important it is for educators to understand how students perceive their experiences, what they consider valuable, and how they define learning. The seminar class gave students the opportunity to question what they learned and for teachers the opportunity to learn that it was crucial to design continuous opportunities for students to have a say in designing their education. As the students in this study reminded us, "Whose school is it anyway?" Without an awareness of ethnicity, how it operates in our society, and how children are situated because of it, it is impossible to create liberating learning environments (Quiroz 2001).

Student views drew teachers' attention to their own assumptions of equity and fairness. While teachers believed that service-learning would lead students to interact with people from different backgrounds and would disrupt stereotypes and prejudices, they neglected to consider the prejudices and stereotypes that students would encounter from other service providers and recipients. Moreover, they learned from students that service-learning sites were not neutral. Although no student denied the value in community service-learning, being viewed as juvenile delinquents attacked their character, educational ability, and left them worried with regard to their future prospects. Their experiences bear witness to the subtle forms of discrimination that occur everyday, even without malice or intention to harm on the part of perpetrators. Connolly and Keenan (2002), argue that in order to resist such subtle forms of discrimination, we need to pay attention to the "effects" such acts have on those subject to them. It is only when we examine "effects" that we can look at unconscious, unwitting, or even good intentions that perpetrate injurious acts and the broader structures and institutions implicated in supporting racism. Teachers also learned that their work needed to extend to the community to help fight negative perceptions of youth and promote a positive image of the school.

CREATING EDUCATIONAL CHANGE

The second data excerpt presented in this chapter gives an account of students "doing democracy." One of the goals of service-learning as it is articulated by advocates of critical pedagogy has been to encourage political learning and civic engagement through service-learning. Research has shown that when students think they have a voice, they are more likely to take ownership of projects and to consider service-learning useful (Wade 1997). By empowering student voices, teachers moved beyond routine reflection on service-learning experiences to facilitate students in being active agents of educational change.

Teachers in this study found that an intentional pedagogical effort is required to respond to students. Creating the seminar and by facilitating the process by which the name of the school could be changed, teachers made intentional efforts to support students' ideas of change.

Students too, learned an important lesson about change and the ways in which change can occur if one is true to the democratic process. Larry, the seminar teacher, in responding to student voices, did not hesitate to challenge them to stay true to the democratic process. Students learned a valuable lesson while being empowered as change agents. They learned the difficulty of dealing with situations where their personal desires conflicted with the process they had adopted. The seminar students wanted to name the school The Innovative School, a name that received few votes from the juniors. They also wrangled among themselves accusing each other of trying to influence the process by lobbying for names they personally favored. Some students wanted to "shred the survey," but later backed away and stayed with the process they had adopted. The seminar gave the students the chance to get involved in a process of change that mattered to them and learn the difficult task of "doing democracy." Larry, the teacher, negotiated a fine line between stepping in to intervene and challenge students and stepping back and letting students have the space for creative dissent.

SERVICE-LEARNING EXPERIENCES FOR A DIVERSE STUDENT BODY

It is clear from the data that good intentions on the part of teachers are not enough to design appropriate service-learning experiences. Teachers in this study learned that intentionality in planning the different components is crucial (Rhoads 1997). Additionally, incorporating student views and ideas both during the planning phase as well as after the service-learning is completed would ensure that the activities planned matched the student learning needs at the levels of knowledge, disposition, and skills. They thought that it was also important to ask questions when designing a service-learning experience—whom will this serve? Whom will this action benefit? What stereotypes are prevalent in this activity? How is it possible to disrupt those successfully? Additionally, teachers began to figure out what types of activities would enhance social visibility in a positive sense so that there were identity related benefits for students of color.

IMPLICATIONS

Critical Multiculturalism in Service-Learning

This study stresses the need to situate service-learning within a framework of critical multiculturalism. As the student experiences in this study indicate, there is a widespread assumption around race and class regarding who serves, who gets served, and who gets seen as productive, contributing citizens in society. A critical multicultural framework may be useful in challenging the widespread perception of service-learning as a White dominated movement. One way to do this is to reform the service-learning knowledge base and acknowledge that service has been a core value in communities of color. For example, the National Indian Youth Leadership Project has demonstrated that service-learning is grounded in the methods of learning and education of indigenous peoples (Hall 1991). Including such multiethnic perspectives in the service-learning literature will move us away from a "helping" model of service-learning to a model grounded on reciprocal power relationships.

Creating Spaces of Possibility

According to Fine and Weis (2003), to pose a critique of silencing in schools requires one to make a parallel commitment to nurturing student voices. The common practice in schools is to "answer questions rather than question answers" (Shor 1996). While service-learning incorporates reflection, doing so within a critical multicultural framework ensures that such acts of reflection are meaningful and hopeful. Seminar classes such as the one described in this chapter offer the space for "extraordinary conversations" that bring to life what may otherwise remain hidden under the surface of routine exercises in reflection. It is possible, as the data in this chapter bear out, for students to do race work of a kind that contributes to dismantling structures and dispositions that support racism in society. Such spaces can also be conceived of as healing spaces or as Roulleau-Berger (1993) refers to them, as reconstructive or creative spaces where students suffering the hidden injuries inflicted by school and society can recover and negotiate a positive sense of identity.

Listening to Student Voices

To affirm that students' perspectives are central to the educating process is perhaps the most important lesson we can learn about student engagement. A curriculum bereft of students' lived experiences is common in schools as teachers and administrators are anxious to get on with the business of learning. The resulting lifeless curriculum is dangerous to already disengaged students or those ambivalent about school and its utility. The energy in the seminar classroom attests to the possibility of creating dynamic spaces where students' experiences breathe life into the curriculum.

This study revealed that empowering student voice in service-learning could lead to favorable results. While scholars have advocated for student voice in service-learning, there have been few examples of what this might look like if framed within a critical multicultural perspective. This study provides one such example where students are able to see themselves as capable of effecting change, a perception that allows them to imagine themselves as adult citizens who can affect policies within a framework of social justice (Youniss and Yates 1999).

Challenging "Sacred Cows"

Promoting student voices within a framework of critical multiculturalism offers more than mere affirmation of students, it offers them a support structure where they can challenge some of the taken-for-granted assumptions around race, class, and education. A critical multicultural framework offers service-learning the means to challenge what Nieto (1994b) calls "sacred cows" in education, or those ideas deemed progressive or ideologically correct. This means that even when educators think that they are designing curricula that are fair and equitable, it is important to question whether such assumptions of fairness, equality, and empowering strategies are indeed benefiting all students (Darder 2002). Similar to Delpit (1994) who made a powerful case for not abandoning skills in favor of the process approach to literacy by arguing that disempowered youth might benefit from the skills approach; teachers in this study learned that it is not enough to interrogate one's motives to assure equity in educational experiences. It is important to consider the effects or the traces that educational experiences leave on students and to examine whether they result in any hidden injuries. In the study reported in this chapter, teachers found that it was important to design service-learning differently for different students so that the experiences they designed did not inadvertently disempower some students while empowering others. If the real work of service-learning ought to be about transcending labels and stereotypes (Claus and Ogden 1999), a framework of critical multiculturalism will help move service-learning toward the twin goals of civic engagement and social justice.

CONCLUSION

In this study, I wished to investigate what takes place when spaces are opened up for students and teachers to have a dialogue on issues that are important to the lives of students. Student voices in this chapter reveal that the site of service-learning informed by critical pedagogy has much to teach educators about what is important to students' lives within such spaces. Such spaces of service-learning and reflection, allow for Fine's and Weis' (2003) "extraordinary conversations," opening up the possibility for new narrations of life experience. Service-learning that centralizes student voices and intersects with critical multiculturalism, opens up a dynamic space of hope and possibility within schools for creating an education for social justice.

NOTES

1. The Community School is a pseudonym as are the names of people and places mentioned in this article. In keeping with research protocol, pseudonyms are used to protect their identities. However, the school does have community service as its main focus and the term community does appear in its name. I have received permission from the school and the participants to use this identifier despite the risk of losing anonymity.
2. At this point, I too got up to leave when one of the students caught my eye and said loudly "you won't want to miss this." Other students glanced in my direction and Carlos and Maria said—"you can stay. You need to know this."
3. In this discussion, the students were referring to community service programs in lieu of jail time. Such community service programs began in the United States with female traffic offenders in Alameda County, California in 1966, and local initiatives followed in several counties thereafter throughout the United States. Nonviolent offenders who would have otherwise gone to prison were given the opportunity to provide community service or make restitution to their victims in lieu of incarceration.

References

Appiah, K. (1992). *In my father's house: Africa in the philosophy of culture*. New York: Oxford University Press.

Barndt, M. & McNally, J. (2001). The return to separate and unequal. *Rethinking Schools*, Spring: 4–5.

Bartolome, L. & Macedo, D. (1997). Dancing with bigotry: The poisoning of racial and ethnic identities. *Harvard University Review*, 67: 222–246.

Claus, J. & Ogden, C. (1999). *Service learning for youth empowerment and social change*. New York: Peter Lang.

Connolly, P. & Keenan, M. (2002). Racist harassment in the white hinterlands: The experiences of minority ethnic children and parents in schools in Northern Ireland. *British Journal of Sociology of Education*, 23(3): 341–356.

Darder, A. (2002). *Reinventing Paulo Freire: A pedagogy of love*. Boulder, CO: Westview Press.

Davis, B. & Harre, R. (1990). Positioning: The discursive production of selves. *Journal for the Theory of Social Behavior*, 20(1): 43–63.

Delpit, L. (1994). *Other people's children: Cultural conflict in the classroom*. New York: New Press.

Dolby, N. (2000). The shifting ground of race: The role of taste in youth's production of identities. *Race, Ethnicity and Education*, 3(1): 7–23.

Fine, M. (1991). *Framing dropouts: Notes on the politics of an urban public high school*. New York: SUNY Press.

Fine, M. & Weis, L. (2003). *Silenced voices and extraordinary conversations: Re-imagining schools*. New York: Teachers College Press.

Gay, G. (1995). Mirror images on common issues: Parallels between multicultural education and critical pedagogy. In C. E. Sleeter & P. L. McLaren (Eds.), *Multicultural education, critical pedagogy and the politics of difference*. Albany, New York: SUNY Press.

Giroux, H. (2003). Racial injustice and disposable youth in the age of zero tolerance. *International Journal of Qualitative Studies in Education*, 16(4): 553–566.

Greene, M. (1986). In search of a critical pedagogy. *Harvard Educational Review*, 56(4): 427–441.

Hall, M. (1991). Gadugi: A model of service learning for Native American communities. *Phi Delta Kappan*, June: 755–757.

LeCompte, M. & Schensul, J. (1999). *Analyzing and interpreting ethnographic data*. California: Altamira Press.

Meier, D. (1995). *The power of their ideas: Lessons for America from a small school in Harlem*. Boston: Beacon Press.

Melchior, A. (2000). Service learning at your service. *Education Digest*, 66(2): 26–32.

Morgan, W. & Streb, M. (2001). Building citizenship: How student voice in service learning develops civic values. *Social Science Quarterly*, 82(1): 154–170.

Nieto, S. (1994a). Lessons from students on creating a chance to dream. *Harvard Educational Review*, 64(4): 392–426.

Nieto, S. (1994b). From Brown heroes and holidays to assimilationist agendas: Reconsidering the critiques of multicultural education. In C. E. Sleeter & P. L. McLaren (Eds.), *Multicultural education, critical pedagogy and the politics of difference*. Albany, NY: SUNY Press.

Phelan, P., Davidson, A. L., & Cao, H. T. (1992). Speaking up: Student perspectives on school. *Phi Delta Kappan* (May): 695–704.

Purpel, D. E. (1999). *Moral outrage in education.* New York: Peter Lang.

Quiroz, P. A. (2001). The silencing of Latino student "voice": Puerto Rican and Mexican narratives in eighth grade and high school. *Anthropology and Education Quarterly,* 32(3): 326–349.

Rhoads, R. (1997). *Community service and higher learning: Explorations of the caring self.* New York: SUNY Press.

Roschelle, A. R. (1997). *No more kin: Exploring race, class and gender in family networks.* Thousand Oaks, CA: Sage.

Roulleau-Berger, L. (1993). Social construction of intermediate spaces. Poor people and social policies. *Societes Contemporaines,* 14/15: 191–209.

Rudduck, J., Chaplain, R., & Wallace, G. (1996). *School improvement: What can Pupils tell us?* London: Fulton.

Ruiz, Richard. (1997). The empowerment of language minority students. In A. Darder, R. D. Torres, & H. Gutierrez (Eds.), *Latinos and education* (pp. 201–328). New York: Routledge.

Shor, I. (1996). *When students have power: Negotiating authority in a critical pedagogy.* Chicago, IL: University of Chicago Press.

Sleeter, C. E. & McLaren, P. L. (1995). Introduction: Exploring connections to build a critical multiculturalism. In C. E. Sleeter & P. L. McLaren (Eds.), *Multicultural education, critical pedagogy and the politics of difference.* Albany, NY: SUNY Press.

Soohoo, S. (1993). Students as partners in research and restructuring schools. *The Educational Forum,* 57: 386–392.

Tatum, B. (1997). *Why are all the Black kids sitting together in the cafeteria?* New York: Basic Books.

Varlotta, L. (1997). Confronting consensus: Investigating the philosophies that have informed service learning communities. *Educational Theory,* 47(4): 453–477.

Wade, R. (1997). *Community service learning: A guide to including service in the public school curriculum.* Albany, New York: SUNY Press.

Youniss, J. & Yates, M. (1999). Youth service and moral-civic identity: A case for everyday morality. *Educational Psychology Review,* 11(4): 361–376.

3

"I Can Never Turn My Back On That": Liminality and the Impact of Class on Service-Learning Experience

Sue Ellen Henry

Service-learning is often framed as a pedagogical perspective and instructional tool that can help "privileged" students gain greater insight into the life experience and perspectives of "others," namely those "served" in the service-learning arrangement. Central to this positive conception of service-learning is a binary between "privileged server" and "underprivileged recipient" or an "us/them" dichotomy. Recently, this dichotomy has been questioned by some researchers as problematic to a transformative understanding of service-learning (Flower 2002; Hourigan 1998; Novek 2000).

Central to the critique of the binary of "server/served" is its overly simplistic approach to understanding those involved in the service-learning relationship. Rather than seeing the complexity and multipositional points of view from which people in service-learning relationships operate, this dichotomy remains too blunt to reveal the variety of identities that both "servers" and the "served" actually live within. While this dichotomy might be rhetorically helpful for those arranging service-learning assignments, it may significantly mask the full identities of those participating in a service-learning experience. What does it mean to the "privileged student server" to share characteristics with the "underprivileged served" service-learning community? How do students, who occupy both privileged and underprivileged status, understand themselves and their multiple identity categories through working in a service-learning situation that puts them in the position to "serve" communities that represent their backgrounds prior to college?

These questions arose for me in my teaching of a multiculturalism course that uses service-learning as a philosophy and pedagogy. Given the local context of my university, most of the service-learning assignments occur in rural, poor, and predominantly white locations. It occurred to me one semester when in this course, three of the 25 students enrolled were from

Central Pennsylvania and were first generation college students. I wondered aloud with them through their service-learning journals and in conversation about the effects of performing "service" in communities that looked exceptionally similar to the home environments they had grown up in prior to coming to college. How was the act of "serving" a local community, composed of people they perceived to be like them, influencing their sense of self? Specifically, how did these interactions influence their identities relative to privilege and lack of privilege?

The service-learning literature has a strong focus on the impact of service-learning on racial prejudice and "unlearning" racism (Boyle-Baise and Kilbane 2000; Green 2001). One problem with the current literature, however, is that "servers" and the "served" most often differ on both racial and class characteristics, with "servers" being predominantly privileged, white college students and the "served" being underprivileged, poor, people of color. In this arrangement, most studies focus on the influence of racial differences between "servers" and "served" and tend to collapse race and class together, making it difficult to determine the influence of service-learning on class awareness, specifically. The fact that "class" is a highly contentious term among sociologists may explain some of this omission. Some sociologists contend that the lack of identifiable lifestyles or value preferences that correlate with income levels renders class status positions an inadequate means by which to understand life in the United States (Kingston 2000). Others maintain that class does have significant influence on the ways in which individuals experience their daily life, and thus warrants distinct and deep study (Aronowitz 2003; Jackman and Jackman 1983; Zweig 2000).

Deliberate study of class in the service-learning literature could be especially productive in advancing our understandings of the ways in which class operates in the United States. The power of the present study emerges from the fact that class is a particularly interesting vantage point from which to see the intricacies of polypositional identity. According to Zweig (2000), the multiple class positions in the United States are shaped through social interactions. Many of us occupy many different classes and operate from various class positions in various life contexts. Service-learning is a powerful tool in that it offers the chance to work with others from different class backgrounds, thus allowing individuals to reflect on their own personal class orientations. As Zweig (2000) suggests, class is a messy category to discuss, but important nonetheless:

> Class is not a box that we "fit" into, or not, depending on our own personal attributes. Classes are not isolated and self-contained. What class we are depends upon the role we play, as it relates to what others do, in the complicated process in which goods and services are made. These roles carry with them different degrees of income and status, but their most fundamental feature is the different degrees of power each has. (11)

This chapter critiques the simplistic nature of the server/served dichotomy, and suggests that we can learn more about the power of service-learning if we look closely at the complex class context that surrounds service-learning work, and the learning that might be taking place for students engaged in service-learning. To support this analysis, I present data I have gathered on a small group of first generation college students attending Bucknell University, who within the last two years, have participated in service-learning assignments in their hometowns or towns very similar to their own. I chose first generation college students as a way of operationalizing working-class status at Bucknell; though not perfect, research suggests that there is a relationship between access to power and income (Aronowitz 2003; Zweig 2000). Working-class status was an important way of matching students with service-learning assignments in the local community in order to isolate class as an element of study. I analyze this data to see what influences the service-learning experiences have on such students' sense of privilege and lack of privilege. This analysis suggests that by complicating the foundational notion of server/served, we cultivate more sophisticated thinking in students relative to their own identities and the identities of others, as well as foster greater empathy and feelings of commonality between groups who might otherwise conclude that they share few similar features or goals. Another element of the analysis suggests that false dichotomies such as "server/served" mask the identities of students involved in service-learning, and has some detrimental implications for what practitioners think about service-learning. Changing our notions of the assumed binaries that undergird service-learning work and expanding our understanding of the multiple identities of both "servers" and the "served" are key to collective and collaborative problem solving aimed at some of the nation's most difficult public problems.

METHODOLOGY

To inform my understanding of the ways in which service-learning influenced the class identity of these students, I selected a qualitative research design. This design relied heavily on individual interviews; spread over the course of two months, I interviewed each student a minimum of three times, with each session lasting approximately one and a half to two hours. Each interview was taped and transcribed, and given to each participant so that she could modify her comments as she saw fit (no students deleted remarks; all their written comments clarified previous statements or the chronological order of events). To prepare for subsequent interviews, I reviewed each prior transcript and created questions to investigate theoretical or thematic trajectories I saw in the data (a process commonly understood in the research literature as member checking). The participants' answers to these questions then became part of the official record of the current day's interview.

In addition to these interviews, I developed a questionnaire that each participant completed. The survey included basic questions such as hometown demographic information, education of parents, estimated annual family income, and plans for the future. From these questionnaires, I was able to develop some specific questions for each participant to gain a greater understanding of her particular family and class background.

I began to review these data sources as soon as they became available, watching for consistent, persistent themes that arose between participants. I then turned to the theoretical literature to see how the comments from these women squared with some of the important scholarship on class and service-learning.

Understanding these women's words also required understanding the college setting in which these women were working and learning. My own experience working at Bucknell for the past eight years helped me to shape specific questions about being working class at a university like Bucknell. In the next section, I share with readers some of this interior knowledge in order to frame the context from which these students' important words arise.

UNIVERSITY CONTEXT AND THE "TYPICAL" BUCKNELL STUDENT

Bucknell University is located in a largely rural region of Central Pennsylvania. Enrolling approximately 3,500 undergraduate students, Bucknell is a private, highly selective liberal arts university with a college of Engineering that attracts a national and international student body. In the class of 2008, 22 percent of admitted students were from Pennsylvania; the remaining 78 percent came from 48 states and represented 32 different countries. Approximately 85 percent of the undergraduates are white; the remaining 15 percent are classified as students of color, including international students. According to the Financial Aid office, 45 percent of Bucknell undergraduates receive some form of need-based financial aid; nearly 65 percent of the undergraduate population receives some form of financial assistance (which includes significant scholarships for athletes on varsity football and men's and women's basketball). The average SAT score for the class of 2008 was 1351, and 93 percent of the class ranked in the top 20 percent of their high school graduating class (Bucknell Admissions Website, accessed April 19, 2004). Tuition at Bucknell in 2004 was just under $31,000, with room and board an additional $6500.

From my point of view as a 39-year-old, White, state-university-educated female faculty member, mother of three, who has worked at the University for eight years, much of the student culture centers around consumerism. Students attending the university tend to be of middle class or more financially privileged backgrounds, with only 12.5 percent of the class of 2008 identifying themselves as first generation college students (Cooperative Institutional Research Program questionnaire, Bucknell University, 2004). Despite the data from the Financial Aid Office that would suggest that a majority of undergraduates are working or taking loans to make their college experience possible, the dominant view of being at Bucknell is one of a privileged life. The school's parking lots are largely filled with BMW's and sport-utility vehicles and the view from the centralized mailboxes in the student center includes regular packages from JCrew and Abercrombie and Fitch. An interesting status symbol supporting this notion of a financially privileged student population is the overwhelming presence of the Vera Bradley tote bag, carried by a significant number of women, particularly women involved in Greek-letter organizations. Priced near $100 for a single tote bag, this bag connotes a sense of privilege and access to financial power that interviewees regularly discussed during our conversations. When talking about the primary messages she got about money growing up "working class" in a neighboring community, Amy (not her real name; all names are pseudonyms to protect the identity of study participants) described her family's reaction to the "bag":

> My dad is very like, he would never want me to go without, but like, he doesn't like me just throwing money. Since I have been at Bucknell I have bought three of the Vera Bradley bags and every time he gets so upset . . . he figures, like one should be enough. (Amy 2004)

Curious about the importance of the Vera Bradley bag, I asked Amy what having it meant to her. Her response, similar to the other two students in the study, was compelling, and in some ways deeply instructive about the need to "fit in":

> I guess to me, I hate to say it, it is like Bucknell. This bag is Bucknell. . . . my sophomore year everyone had them, so I wanted to fit in, so I got one too, and I like them. So, I've gotten a couple more. . . . this makes you part of this crowd. This is like wearing a t-shirt that says Bucknell. (Amy 2004)

Jacqueline, who doesn't own a Vera Bradley bag, nonetheless articulates the importance of having such a bag:

> [The bag] definitely gives a sign like, I have this money to spend, and this is how I am going to reward myself. Money doesn't really cost; I want the bag, everyone has it, and I need it too. [But] I just think its overpriced. I don't feel like this is something that I can spend money on for me, but I feel like it is something that I can spend money on for my best friend. (Jacqueline 2004)

These sorts of pressures are not new, per se, but as Horvat and Antonio (1999) elucidate, this pressure to conform and "look the part" can spread into other areas such as language use and particular customs. Amy discussed how her roommates, who she identified as "coming from money" often pointed out her Pennsylvanian "accent" and their frequent corrections of her speech. Amy tells of the conflict she has felt as a member of the Bucknell community and her native home culture:

> [Amy]: I think . . . sometimes I do things, things that get looked down upon. People will say, "oh my gosh, I can't believe you watch NASCAR," or will comment on certain words I say . . . my accent. Apparently I have an accent.
> [SEH]: I can't hear an accent.

[Amy]: I tend to say like "yuens" for "you all" or "everybody" and I really get picked on about that all the time. But I really feel that at times, there is a conflict here. . . . Like when I said to my friends that we could go "sled riding," my friends said it was "sledding." That is what I call it . . . I felt like saying, "don't correct me just because I call it something different." (Amy 2004)

Thus, even though a significant portion of Bucknell undergraduates do receive aid to offset some of the exorbitant cost of tuition, there remains a particular image of the archetypal Bucknell student as financially well-to-do, on the road to being a stock broker or working in the family business. Interestingly, in 2003, 43.6 percent of Bucknell undergraduates reported family incomes of over $100,000; 16 percent of this population reported family incomes of $250,000 or more.

This image of the typical Bucknell student as potentially more interested in clothing choices than social issues was a theme consistent throughout the interviews with these three first generation students. This perspective was clearly articulated as an academic problem by Jacqueline, a recent graduate of Bucknell, who is now in graduate school in Ohio. Discussing a capstone course entitled "Italian Culture in Central Pennsylvania" she took in her last semester at Bucknell, Jacqueline described a sense of disengagement among her more wealthy peers that at times she found distressing and unnerving:

[Jacqueline]: I felt like I was the outsider. Most of [the other students] were "mom and dad paid for all Bucknell's tuition." They were not engaged at all.
[SEH]: How did you know that their parents paid for their tuition?
[Jacqueline]: I am sure of some of it; [some of my information was] assumptions that I made. Based on appearance, based on what I knew about them, based on the fact that you would see them drop $50 every weekend, every weekend night at the bar. It was just a crowd of people that were all friends. That is why they took the class together. They weren't there to be students . . . they want to say, "put on the PowerPoint exactly what I need so I can write down some notes, so it will help me to write my paper." So it will be easier. So I don't have to think about it. . . . It was almost like "my parents are paying for this." (Jacqueline 2004)

Privilege at Bucknell seems to influence not only students' perceptions of one another, but also the dominant view that local residents hold. Bucknell students are commonly assumed by locals to be wealthy, white, and from New Jersey. For example, when Amy began her student teaching assignment in a local school, many of the students assumed she would fit their image of the "typical" Bucknell student:

One of the classes said to me, "oh, did your parents buy you a BMW for Christmas and are you from New Jersey?" (Amy 2004)

In all, there is much to support the idea that there exists intense pressure at Bucknell to look and appear as someone who has access to financial power and a promising career making even more money. Indeed, this image is so powerful that it dominates not only on-campus relationships but has also become the consistent viewpoint of others in surrounding communities.

But, of course, not everyone at Bucknell exists in this world. There are a growing number of working-class students who, through financial aid and local full tuition scholarships, are enrolling at Bucknell. It is important to recognize that access to this type of education is not widely afforded by everyone; at $31,000, the vast majority of high school seniors in Pennsylvania would not be able to afford such an education, even with substantial aid. But while gaining admission to Bucknell and being able to successfully matriculate does connote an important sense of privilege, for many students this is only a partial story of their identity. Nationally, like many of their peers, first generation college students at Bucknell overwhelmingly come from working- and lower middle-class families (Terenzini et al. 1996) and are often geographically constrained in their college selection (College 1995). The average family income of the students in this small study was $49,000; the 2002 median family income in Pennsylvania was $43,577 (Bureau 2002). Median family income for the United States in 2002 was just over $42,000 (Bureau 2003). Median family income for Bucknell students receiving need-based financial aid was $70,000 in 2004.

All three of the students involved in this study are first generation college students. Each is white, female, and from the Central Pennsylvania region that features small towns with largely crumbling economic infrastructures. Katelyn, who earned an undergraduate degree from Bucknell and is now completing a graduate degree in Education at Bucknell, has lived her entire life in a small city 30 minutes north of the university. During the industrial era, this city was known for its logging industry. Since 1950, this industry has left the area, leaving behind a city with little economic support. Jacqueline is from a town 45 minutes east of Bucknell that previously depended largely on the coal mining industry. With the change in fuel sources throughout the nation, the coal industry has since left this area, and in its wake, left thousands of people to make their livelihoods in other ways. The third student, Amy, came to Bucknell from a town she describes as "shabby." She showed me pictures of her neighborhood; each house had a barrel for burning trash, and several of the homes had rusty cars in the yards. All three of the students describe their backgrounds as "working class." When I asked them each to elaborate on how they have come to this conclusion, their responses are remarkably similar and illustrative of significant elements of working-class identity:

> [Amy]: [My parents] just work hourly rate jobs, and they just have their high school educations. . . . My town is a very low income area. The houses are very run down. Junk cars all over the place. . . . A lot of people in my town are like, on welfare.
>
> [Jacqueline]: My parents work hourly-wage jobs and always have, even after many years of working. [My town] is definitely in a lower class area.

I wouldn't say poverty stricken, but definitely the lower levels of middle class and the upper levels of the lower class, socio-economically. . . . I see my parents work very much in jobs that people don't really like to normally do. From the factory work, to stocking shelves in a grocery store and things like that.

[Katelyn]: I would say working class. Even though my father was a driver for UPS and made unbelievable money. He made money, but there was no sharing, no spending of it. We had food, but we weren't allowed to touch it. . . . And he didn't support education. [My father] made me think that living in [my town] was all that there was, that there is nothing else out there. . . . I mean, obviously I knew there are doctors, there are lawyers, there are people that you know go to school, yet I think because I didn't really have an education, I didn't think it could happen for me. Working class for me meant living my life in a small radius, both literally and figuratively.

These descriptions correspond highly with Zweig's (2000) analysis of working-class identity, which emphasizes the power that one has in making economic and occupational decisions. Based on this criteria, Zweig maintains that there are three main classes in the United States: working class, middle class and capitalist class, the largest group being the 60 percent classified as working class. Consistent with a Marxist point of view, Zweig's major criterion for placement in the working-class category is access to power, particularly power at work:

Working class people share a common place in production, where they have relatively little control over the pace or content of their work, and aren't anybody's boss. . . . the recent increase in inequality is not just a case of the rich getting richer and the poor getting poorer, as the media often portray it. Our society's growing inequality of income and wealth is a reflection of the increased *power* of capitalists and the reduced *power* of workers. . . . Class has its foundation in power relations at work, but it is more than that. Class also operates in the larger society: relative power on the economic side of things translates, not perfectly but to a considerable extent, into cultural and political power. These forms of power in turn reinforce, adjust, and help to give meaning to classes.   (Zweig 2000, pp. 3–4)

Using this definition of working class, each of these women meets the criterion and is accurate in her self-assessment of her class position. Their position as members of the working class, however, does not tell the full story of how these women view themselves in society. Indeed, all three of these students noted that attending Bucknell was an illustration of their social privilege. Each of these women came to Bucknell on full scholarship funded by a local philanthropist who offers monetary support to financially underprivileged, highly accomplished local high school students. In the context of their working-class backgrounds, the idea of coming to Bucknell strikes these three students as very special and personally important. As Katelyn and Amy assert:

[Katelyn]: [Being at Bucknell] is almost like a dream, a dream state so to speak. I hung my [undergraduate] diploma [from Bucknell] where I can look at it all the time. [It reminds me that] nobody did [this] for you. And if you were able to do that, you are able to do other things as well.

[Amy]: Being at Bucknell makes me feel really smart. Because, I got in here and I get to go here. I think that when people look at that, they are going to think that about me and say, wow, she must have been pretty smart to go to Bucknell, and graduate from here.

Being working class is a fairly unusual state at Bucknell, given the cost of tuition. However, when these students then go out to their service-learning assignments, they reenter communities extremely similar to their own working-class roots. How would such an experience influence their sense of self-identity?

THE SERVICE-LEARNING ASSIGNMENTS

These three students were part of an elective course I teach entitled "Multiculturalism and Education" during which each student invests a minimum of 15 hours over the semester in a service-learning site of her choice. To support the service-learning component of the course, I prearrange sites with various social agencies such as schools, prisons, and local nonprofit organizations. Each of these students took this course within the last two years and worked in a different service-learning site. Amy spent time teaching and observing in an alternative high school education center that draws its student population from several surrounding counties. Jacqueline worked as a peer mentor with a brother–sister pair of limited-English proficient children from a low-income neighborhood near her hometown. Katelyn volunteered as an English as Second Language tutor and also worked with the primary administrator to offer clerical support for the program at a high school 20 minutes from where she grew up.

SERVICE-LEARNING AND SELF-UNDERSTANDING

Consistent with other research findings, all three of the students involved in this study indicated that they became more aware of their own sense of privilege during the course of their service-learning assignment (Green 2001; Jones and Abes 2004). In particular, these women noted that service-learning assisted them in their growing awareness of racial, language, and class privilege. All three, however, were also aware of moments and incidents that reinforced for them some of the categories that would traditionally make them underprivileged, and thus they frequently noticed overlaps and similarities between themselves and the people they were working with through their service-learning assignment. These complex outcomes influenced these women in their self-understandings, in particular highlighting their liminal class identity at Bucknell. Not only did they frequently feel like outsiders at Bucknell, with its financial and educational elitism, but they also felt like outsiders when being university representatives during their service-learning assignments. This liminal position underscored for these women that their identity constituted a mix of privilege and underprivileged elements and that pieces from both categories were important to their sense of self.

Feelings of Privilege

All three students indicated that through their service-learning experience, they were able to see ways in which their lives had been shaped by "unearned privileges" (McIntosh 1988) of which they were just beginning to be conscious. Privilege associated with native English language ability especially impressed Jacqueline and Katelyn, both of whom worked with limited-English proficient students. As Jacqueline reflected on her time spent with an 11-year-old brother and sister from a local community, she remarked on the important awareness she developed about some previously unexamined beliefs:

> I guess it just really opened this door for me to think about the fact that this [my hometown] is a lot more diverse than I give it credit for being. . . . This is what I know about it now. There are a lot more students who come in [to my hometown area] that are not English speakers. . . . it was hard because being in the area like this, [there are] certain staples in the area that you could talk [about], or you think can talk about with anyone. You can drop the [name of the local amusement park] and people are going to know what you are talking about. . . . You take that for granted. . . . It just opened my mind, like how much I take for granted about the area in which I grew up. . . . I think it has just really opened up my eyes to really say like; here is a very strong difference where you didn't think there was going to be one.   (Jacqueline 2004)

Katelyn spoke extensively about working with limited-English proficient students and what this experience taught her about her own racial and language privilege. Among her many comments were these thoughts about the demographic imperative facing the United States and how this demographic shift influences racial privilege:

> [Working at the high school in the ESL program reminded me of the] reality that whites will become a minority. And as a white person, I would not want to be treated the way I know that minorities are treated. And if I am going to be a minority some day, I hope that somebody would take the time, the effort, the patience, the care with me trying to learn something that I didn't know, the same way I did with those students that were so bright, they were so bright, but just because they didn't understand English, it was almost like they were penalized.   (Katelyn 2004)

Another source of privilege that all three students identified was their educational background. They all recognized that being the "norm" in their home environments had made it substantially easier to have success in their early educational experience, which ultimately led to their being accepted and supported at Bucknell. Contrasting their educational experience with that of their service-learning partners helped them notice significant differences, specifically the dramatic hurdles placed squarely in front of their service-learning partners due to their limited English proficiency, racial minority status, and location in the poor working class. When working with an alternative education program, Amy noted how graduating from an alterative school might hinder one's chances later on:

> [Amy]: I wonder . . . do any of these kids have the chance to go to [Bucknell]? I think probably not because a lot of them stay [at the alternative school] until they graduate. I wondered about that too. Is that really fair? How can they get into college? They graduate from the [alternative school]. I kind of wondered if that was really being fair to them. [The alternative school] was just so different from like the school environment. I don't see how they could ever go back to a [traditional] school environment. . . . [they were always] walking around the room. If they got up and did that, in a [traditional] school environment, the teacher is not going to put up with that. (Amy 2004)

Jacqueline wrote her final research paper for this course on the state of ESL certification in Pennsylvania and the fact that until 2003, the state had no required certification program for teachers working with ESL students. Having been in teacher education for secondary mathematics certification and having a solid understanding of the numerous classes required to gain certification, this situation shocked Jacqueline enormously:

> I had a lot of frustration with the state. . . . I was just blown away by the fact that there was no vigorous training for this. . . . the credentials are so extensive to teach elementary education and math; 856 things you need to do to teach elementary education, math or whatever. But here we are with these students who are already a step behind because they . . . can't communicate with anyone. Why isn't [teacher preparation for ESL] more intense? Why is that not valued? Why are these students not valued more as learners in this state and in this country to say no, we need to train our teachers better? (Jacqueline 2004)

While all three of these students were working in either their hometown or in towns extremely similar and physically close to their hometown, they were just beginning to see some new elements of their home and personal experience. These women saw themselves as privileged to be able to attend Bucknell University. Through their service-learning assignments, they were able to begin to question the meritocracy behind this experience and began to actively explore what it really meant that they were selected to attend Bucknell. Was everyone from their home environment *really* encouraged to have these same experiences? Were all people *really* equal in their ability to go to college, or move from working class to the middle or upper-middle class?

Consistent with other studies, all three of these students soon recognized that they "took for granted" experiences and access that is largely a function of unearned privileges built upon their racial identity, native English facility, and class status. These women had all worked hard in school and had put achievement in school as a top priority; and yet, their service-learning assignments helped them see the systems of support working to aid their success that they had previously ignored. They were also better able to see how others, particularly those disadvantaged by being poor/working class, of limited-English ability, or by being a racial minority (or the combination of these factors) dramatically affected the internal "safety net" they had previously assumed was present for everyone. Jacqueline was particularly influenced by this set of thoughts during her service-learning experience, and like many students in other studies, expressed these thoughts in the form of intense guilt. She discussed at length her realization that the college atmosphere encouraged her to believe that others would take care of teaching her what she needed to know rather than taking responsibility to teach herself for the future:

[SEH]: Would you say that your service-learning opportunity was kind of more emotional and cognitive or more cognitive than emotional?

[Jacqueline]: I don't think you can separate them at all. [My classmate and] I used to drive back from dropping [the brother and sister] off at their home and there would be days that we wouldn't even have words to say anything. There were days that I felt so guilty. Like God, I have so much of this wonderful stuff that I take for granted every day. You get filled with a sense of guilt, like why am I not doing more for those who aren't as fortunate as me. [This was] a good experience, it has really opened me up to like rethink a lot of things that I took for granted. I think it is so easy in college to do that. You are in a sense somewhere you don't have to necessarily think about that stuff because everyone is so focused on you and your growth and your development. [The college environment] can end up making you feel like someone else will teach me what I need to know. . . . I am amazed every day and how much I take for granted.   (Jacqueline 2004)

Feelings of Similarity and Lack of Privilege

While these three women expressed many of the typical responses to working with others in service-learning arrangements, namely feelings of guilt, responsibility, and intense empathy for others' lived situations, they also expressed feelings of similarity and overlap with their service-learning partners. All three of the students indicated that their service-learning site was familiar to them in many ways; the setting was comfortable, accessible, and "known" to them. In particular, those students who worked in school settings described feelings of great familiarity and understanding about their service-learning locations. From the vantage point of privilege, these students began to see how these familiar settings were alienating to others they worked with; how what had largely been good for them was detrimental to others' learning and their futures. Bringing together these feelings of familiarity with the disjointedness of being working class at Bucknell helped each of these women see what they shared in common with their service-learning partners, namely feelings of isolation, being treated poorly by many in the community, and being assumed to be unable to do academic work.

Amy's experience during her service-learning and her student teaching assignment are particularly emblematic of this new level of self-awareness. Amy was assigned to a local school for student teaching, and was assumed by her high school students to be a "typical" Bucknell student. Amy expressed significant discomfort with this assumption and described her response to her classes upon learning of their assumptions about her:

> [Amy]: One of the classes said, oh, did your parents buy you a BMW for Christmas and are you from New Jersey? I was like, no what, no, I live an hour and a half away. I drive a Neon. I just wanted to be like, I am just like you. I could tell, like I was similar and from a similar area to the area that I student taught in. I wanted to be, I am just like you I am not your typical Bucknell student.
>
> [SEH]: What kind of impact did this experience have on your sense of who you are?
>
> [Amy]: It kind of made me feel like, . . . I don't fit in anywhere. Like you see Bucknell students as this, and I am not that. I think [my students] started to realize just by things I said, like we were talking about my father one day in class because we were talking about state jobs. I was like, my father works for the state and it is just a regular job. I didn't say my father, the stock broker, etc. . . . and like NASCAR came up in class, I think they started to realize that I was just like, I was similar to them.
>
> [SEH]: So you saw a change in their assumptions that they had laid on you at the beginning?
>
> [Amy]: Right, yea.
>
> [SEH]: Do you think that that made you more comfortable in the environment?
>
> [Amy]: Yes. . . . I would have not been comfortable teaching at [a wealthier local school]. I have heard the type of students that go there. I wouldn't have been comfortable there. Because I would have been feeling the whole time that they were thinking that they were better than me. I was, just more comfortable teaching students who are like me . . . socially like me. And economically like me. . . . Economically I would be like students whose . . . parents like don't have really high paying jobs. They have jobs but not like extremely high paying, like professors or whatever. Socially, we have a lot of the same interests. NASCAR came up, like one kid wore a Nascar shirt. And I commented, and things like that. Just like the same interests. The same, I feel like I have the same economic statuses as most of the students that I taught. I just felt more comfortable.   (Amy 2004)

Katelyn spoke regularly during our interviews about the feelings of isolation and disconnect that she regularly experienced while being at Bucknell and how seeing these same feelings in the ESL students she worked with influenced her thinking. Katelyn told me her story of being insulted by a faculty member at the technical school she attended before coming to Bucknell, based upon her class and educational background:

> We were discussing the term "hegemony" and I didn't understand what the
> word meant. When I asked him to help me understand it, he said that if I didn't
> know this concept, that when I went to Bucknell, people would think of me as
> "nothing more than a loud mouth cheerleader."   (Katelyn 2004)

When Katelyn went into her service-learning assignment, she felt much
the same types of insults were being communicated to the ESL students she
worked with.

> [Katelyn]: [People at the school] sent messages to these kids, "hey, your skin is
>   different, your color is different, you don't know how to speak English and you
>   don't fit in here. You are below. You don't know anything because you don't
>   know how to speak English." That was really loud and clear at the school.
> [SEH]: Do you remember an example of where you saw this happening?
> [Katelyn]: All around me, everywhere. Everyone, in their interactions with
>   these kids, sent the message "this is it, this is all that is ever going to be for
>   you." . . . Specifically I can remember being inside the [ESL] room and the
>   teacher did not know how to speak Spanish. And he acted very meanly
>   toward the Spanish language like it was something . . . to be made fun of.
>   I can remember this student coming up to him and trying to tell him some-
>   thing, and the student was very upset. This teacher just treated the student,
>   like because the student couldn't express himself in English properly, that he
>   was just pathetic, and said "go talk to the director [of the ESL program, a
>   native Spanish speaker], he will understand what you are saying." And
>   I thought, oh my gosh, I could see the look on that student's face when he
>   walked away. I could see the look of just one more little piece being taken
>   out of him, of making him feel like he didn't have the right to be alive. He
>   looked like he felt "I just don't deserve to be human."   (Katelyn 2004)

In Amy's case, she was very much like the students she worked with during
student teaching and in her service-learning assignment on some important
cultural markers such as race and class. Katelyn's situation is a bit different
because she didn't share these features with her direct service-learning
population. In both cases, however, both students were able to connect with
their service-learning partners through their experience of class; for Amy,
class was the feature she wanted to use to connect with her students during
student teaching. For Katelyn, it was her experience being the target of a
class insult that fueled her empathy and feelings of connection between
herself and her ESL service-learning partners.

   This multipositional understanding of personal identity was also an
outcome for Jacqueline, who described how her service-learning experience
increased her interest in working on behalf of others, particularly people
who are typically overlooked in college. As a first generation college student, Jacqueline has significant insight into how one might feel as an
outsider; while privileged in her ability to go to college, she understands that
among those in such a privileged environment are many who are having

trouble fitting in and finding their place in the university. During our interviews, Jacqueline described her feelings about coming to Bucknell and some of the early messages the school sent to her about how she might change as a result of her university experience:

> The handbook. You open that first flap in the handbook and it said "you can be anyone you want to be here." Like, flat out, "try a new personality, try a new this, try a new that," and that was appalling to me. I am not coming to this school and completely hide who I am. Growing up in [my hometown] was very important to me, and it wasn't something I was going to let Bucknell take away. I wasn't gonna hide from people. I like growing up where I came from. I like the coal region. And some people indicated to me, others from my town and the larger population at Bucknell, "who would want that? Why would you tell people [you were from the coal region] when you didn't have to?" (Jacqueline 2004)

During her service-learning assignment, Jacqueline saw similar messages being sent to the limited English-proficient brother–sister pair with whom she worked. As mentioned earlier, Jacqueline was deeply troubled by the lack of educational support that her service-learning partners received. Without the benefit of trained teachers, and in her specific case, without having a dedicated ESL teacher on site at the elementary school her brother–sister team attended, Jacqueline rightly questioned the degree to which these children could benefit from their educational experience. The message she determined was being sent to these children was that they weren't worth educating; that their natural makeup as Spanish-speaking brown children rendered them unimportant to the school system. Being part of this experience with her service-learning partners helped her understand the connection between her own experience of not being valued for what she brought to school and the ways in which her service-learning partners were treated by their school. Having a visceral understanding of this injustice galvanized Jacqueline's commitment to advocating on others' behalf. During her interview, Jacqueline revealed her reasons for going to graduate school and how her service-learning assignment influenced this choice:

> . . . your class opened my eyes up so much that I had never thought about. Social justice was never something that I would put as a top value. But as I keep going through this experience, [I keep thinking] how can I make this better for those students that no one cares about? And it could be as simple as when sorority recruitment goes on in my building; no one cares at all about those women who are not Greek. . . . I was like, I am letting hundreds of women walk through [my residence hall for sorority rush] and completely disrespect the lives of those students who are not Greek. . . . A lot of it started with your class opening [and the service-learning placement] . . . [it pushes] these students even further on the margins.   (Jacqueline 2004)

What emerges from these insights is that these students, through their acts in service-learning, not only see themselves as privileged due largely to their educational background, but are also able to connect with their service-learning partners through reflecting on elements of themselves that are less-than-privileged. In other words, these students experience their privileged/underprivileged components as a dialectic, a set of positions that they move in and out of constantly. All three recognized the importance of their college education as a source of privilege; all three saw the privilege in their own lives, whether it be due to their education, family support, racial identity, or English-language ability. They also, however, understood that they shared some important characteristics with those with whom they worked in their service-learning situations. This sharing of similarity was an important catalyst for their own thinking about themselves, and in particular their understanding of what they may have to offer particular groups who are often forgotten.

CLASS LIMINALITY AND IMPLICATIONS FOR SERVICE-LEARNING

These themes have serious implications for the traditional server/served dichotomy in the service-learning literature. The problem in this dualism rests on an untenable assertion that the positions of "served" and "server" are truly distinct from one another. In fact, what these three women point out is that, when it comes to human beings, the positions of "server" and "served" are matters of situational degree versus being fixed positions in the world. Particularly damaging about the either/or set up of the server/served dichotomy is the dilemma it creates for individuals. As Sonia Nieto discusses in her essay "On Becoming American: An Exploratory Essay," second-language learners in the United States are faced with the forced choice of distancing themselves from their native culture in order to become "American": "either lose your 'private language' to become a public person . . . or retain your 'private language' and forfeit your public identity" (Neito 1998, p. 50). Dichotomies such as English *or* Spanish create systems in which second-language speakers fail. As Nieto suggests, a better solution is to look to systems that support "English plus Spanish" (Neito 1998, p. 50).

Similarly, by reifying the notions of "server" and "served," we run the risk of supporting the very systemic distinctions among people that service-learning work is meant to explode. Leaving college students in the traditional position of "server" carries the great likelihood that students will continue to see themselves primarily as privileged persons amply able to give service, rather than as a partner with others who share important feelings, beliefs, motivations, and values. Rather than working with others to unearth and address systemic problems in achieving social justice, charity becomes the galvanizing force behind one's community contribution under the "server/served" dichotomy.

Another interesting outcome of this small study is the influence of class and service-learning on the individual student participants. In many ways, these women were not part of Bucknell because they did not fit the stereotypical profile of financial wealth and elite educational preparation. Being working class—having to work during college and in the summers to make money (rather than travel for the summer or just focus on schoolwork during the academic year), returning home during spring break (rather than travel to another country or take a cruise with college friends), driving an older American-made car, if one is driving at all (rather than driving a recent model import or SUV)—made these women's experience at Bucknell different from many of their peers. Impressive is the fact that these women acknowledge that their experience is different from the typical Bucknellian; Katelyn's earlier statement about the Vera Bradley tote bag being "Bucknell," and Jacqueline's remark about the tote bag being a highly valued item even for herself (though she states she could never bring herself to buy one for her own use, only as a gift for someone else) suggest the distance between who these women are and the values of the dominant peer group around them. In some ways that were *at* Bucknell but not *in* Bucknell. They all valued their college experience and believed that they earned a superior education and made good friends. And yet, there is also a certain distance that seems apparent in their comments; it is as if they are being pulled in two directions at once.

If social mobility is the road that arises from the Bucknell experience, remaining true to your family and hometown values might be considered the other road to travel. As Jacqueline's comments about the university hand-book reveal, she immediately felt the university calling her to make a choice between who she was growing up and who she was going to be as a Bucknell student. bell hooks writes eloquently about this dilemma when describing her experience as a working-class black woman encountering elite higher education as a graduate student:

> Throughout my graduate student years, I was told again and again that I lacked the proper decorum of a graduate student, that I did not understand my place. Slowly I began to understand fully that there was no place in academe for folks from working-class backgrounds who did not wish to leave the past behind. That was the price of the ticket. Poor students would be welcome at the best institutions of higher learning only if they were willing to surrender memory, to forget the past and claim the assimilated present as the only worthwhile and meaningful reality.   (hooks 2000, p. 36–37)

Supportive of these powerful, personal arguments is an empirical research study on working-class academics conducted by Jack Ryan and Charles Sackrey (Ryan and Sackrey 1996). In this work, these authors explore the responses university professors from working-class backgrounds have to their new class status among the educational elite. Central to their analysis is the notion that coming from one class background and moving into another is unilaterally challenging, even if the individual chooses to completely assim-ilate to the customs, language, and culture of their newfound class destination. As Lucile Duberman points out, such social mobility, especially in a society like the United States, is likely to cause discomfort for people such as bell hooks and the three students who participated in this study:

> The mobility experience in a status-minded society is likely to have some disruptive consequences, either because of the status orientation or anxiety of the mobile individual, or because of his inability to adjust successfully to the new groups into which he moves, whether up or down. (Duberman, cited in Ryan and Sackrey 1996, p. 114)

This tension is very likely to occur for those in the teaching profession, in part because members of this group change as a result of their higher education, yet frequently return to work with communities similar to those in which they grew up. Ryan and Sackrey (1996) quote Eric Olin Wright on this matter, who confesses that to be a member of the teaching profession is "to be objectively torn between two classes" (Eric Olin Wright, cited in Ryan and Sackrey 1996, p. 107). Expanding their understanding of Wright's acknowledgment, Ryan and Sackrey suggest that "the academic work process is essentially antagonistic to the working class, and academics, for the most part, live in a different world of culture, different in ways that also make it antagonistic to working-class life" (1996, p. 107). This antagonism is well documented in the sociological literature, dating back to the late 1950s (Blau and Duncan 1967; Lipset and Bendix 1959; Duberman 1976; Coleman and Rainwater 1978).

Such antagonism, according to Ryan and Sackrey, arises from the "sense of being nowhere at home" that leads to "internalized class conflict" among working-class academics (Ryan and Sackrey 1996, p. 113). In response to these experiences, Ryan and Sackrey found four primary orientations among working-class academics to counter these often deeply troubling circumstances of their social class ascendancy: acceptance, separate pathways, balancing class locations, and outsiders. Among those who practiced acceptance, the dominant notions were to learn the rules of the new environment and work within the newfound framework. Those operating from a separate pathways orientation worked to find new ways to be a professor within a largely static system without allowing their "large disenchantment with the university" to devolve into cynicism (Ryan and Sackrey 1996, p. 161). Balancing class locations demarcates those who are working to bring together with integrity the dueling forces of desire for social mobility with "loyalty to one's friends and folks" (Ryan and Sackrey 1996, p. 197). People working from this vantage point describe the enormous effort it takes to remain true to both roads, rather than aiming toward "acceptance" or forging "separate pathways." Finally, those working from an outsiders position operate figuratively and at times literally "outside" the university.

Even though Ryan and Sackrey's findings arise from an adult population of highly educated individuals, it is interesting to apply this framework to the information gathered from the three first generation college students who participated in this current study. In large part, it appears that the three women in this study are working at balancing their class locations, at trying to bring together a sense of their class of origin with their hoped-for class destination. Amy's comments about her sense of comfort student teaching in her working-class high school, her desire to tell the high school students early on that she wasn't a "typical Bucknell student," and her sharing of important features with the students such as an interest in NASCAR and parental occupation is perhaps the strongest evidence to support this contention. And yet, Amy also suggests that going to college was a calculated choice to gain social mobility, at least with regard to educational attainment:

> [SEH]: So [when you decided on going to college] were you wanting some upward mobility?
> [Amy]: Yea. But maybe on just some of these other factors, like occupation, like actual personal income. Some of the stuff that you own. [But there are also] activities you want to kind of hang on to that you grew up with.   (Amy 2004)

Another example of the importance of balancing class locations among the participants is seen in Jacqueline's description of the "transformation" into someone more "typically Bucknell" by a high school classmate who came to Bucknell at the same time she did.

> [Jacqueline]: I remember this very distinctly because I went to Bucknell with two other girls from my high school class. We were in the same residence hall our first year. . . . I started to see this girl start transforming into a completely different person than she was in high school. In a month it was like, who is this girl? She is not who we hung out with in high school.
> [SEH]: And what was different?
> [Jacqueline]: Who she was. 100% of who she was and the things that she got involved in and what she did was not like in high school. You wouldn't see her at a social event ever [in high school], and then that all turned off, and she would say [in college] "this fraternity party I went to, and this thing I did and this other thing I did." [I wondered] who are you? (Jacqueline 2004)

Perhaps most revealing in this description is Jacqueline's incredulity at the way in which her high school classmate handled what she may have experienced as a serious dilemma, and perhaps even the same dilemma of mismatched class positioning that Jacqueline faced as a first generation college student at Bucknell. Jacqueline's disdain for her friend's "acceptance" approach is coupled with her own attempts at balancing her class positions. Her work to make her working-class position evident in her life can be seen in her strong words of pride for her hometown. Her focus on social justice can also be understood as an operationalization of working-class values aimed at the disenfranchised in her particular environment. In her service-learning experience, it was those who were second-language learners who needed Jacqueline's outrage. In her current context, it is the women in her residence hall who are not participating in sorority rush who are marginalized and require Jacqueline's advocacy.

Applying a dialectical approach to understanding the class position of students in service-learning relationships describes the reality of the participants in service-learning better than the traditional dualism. As we can see from their words, Katelyn, Jacqueline, and Amy may have first come to their work in service-learning from the position of "privileged college student server," but the real education of the service-learning assignment came from their reflections on what they shared with their service-learning partners. In fact, their identity—who they knew themselves to be and who they wanted to become in the future—largely arose from the fact that they shared some important characteristics with the service-learning site, namely a similar class-background and feelings of isolation and lack of personal value. While each of them left their service-learning assignment with an understanding of the privileged aspects of their lives (language, race, class status) they also began to understand the underprivileged aspects of their lives as well and how these two components came together to make up the persons they are today, the persons they are becoming.

If we envision and reify college students as "privileged" servers and those they work with as "underprivileged served" we fail to see students in their totality, especially how their class status changes and is influenced by the relationships they make during their service-learning assignment. At Bucknell, these first generation college students' families were far less wealthy than many of their peers and, as we have seen, their accents and other elements of their home lifestyles were critiqued and denigrated by other students. Yet, when these women went out to their service-learning sites, they were seen primarily as Bucknell students and all that this label connoted for local people. Due in large part to the relational nature of class (Zweig 2000), working in their service-learning capacity afforded these women an experience of being able to construct their identities, while understanding that these constructions are also socially based. Being at Bucknell suggested to these women that they were on the track of upward social mobility. Working in their service-learning situation taught these women that there was much that they liked about themselves from their working-class background that they wanted to maintain as important elements of their identity. The combination of these powerful forces also served as the basis for their work with their service-learning partners and their future career choices.

One example of this linkage was Katelyn, who, after exploring her working class background and largely racist upbringing, explained that respect was a central value in her family. Her work in the course helped her realize that in her family respect was a quality extended only to white people. And yet, this early focus on respect, while altered from the way in which she was raised, culminated into a core belief she brought to her graduate school education and her future career in higher education administration:

> I gained a great set of values and morals and ethics and manner and a lot of self-respect and respect for others came out of [my] upbringing. [When deciding to

leave an abusive relationship] it was respect, respect for myself and others, that motivated me, not only respect for certain people [as my parents had]. This is what I bring to my work now; respect for me [now] is for all, everyone, not just [depending] on what color you are. (Katelyn 2004)

Exuding pride in her hometown experience, Jacqueline explains that it's connection to people that she holds central from her core working-class values:

[My hometown] was the best place for me to grow up and shape who I am. I can never turn my back on that. I felt like I had a lot of good things growing up that a lot of people that I met at Bucknell did not have. . . . [during college] when I talked about high school and said things like, our teachers came to our games or that we had good relationships with our teachers and they were interested in our personal lives, people looked at me like I was nuts. Hanging out with your teachers, why? You actually shared your personal life with your coach? Why? How? (Jacqueline 2004)

Now, as Jacqueline asserts, being supportive and staying connected is an important value she draws on in her current work: "[now], it is like you can have your opinions and we can have ours, too, and we may argue about it, but we will try to understand where each other are coming from. . . . I think that the value of being supportive of one another even though we might differ is definitely [something I take with me]" (Jacqueline 2004).

As a teacher committed to service-learning, I take away a great deal from the conversations with these three students. Helping students understand their pasts differently, specifically in light of their new education that emerges from their service-learning assignment, is now a significant part of my pedagogy. This focus requires that I work to not see my students from the "typical" Bucknell perspective; it calls for me to remain open and questioning with the members of this course (and my other courses, as well) about what their backgrounds taught them about themselves and others, and then ask them, how is what you knew to be true from these lessons different since your work with your service-learning partner? How are you different? How are you the same, or potentially, *even more* of who you were than before? Such questions, I believe, engender the notion that many of the identity markers we use to describe ourselves are fluid; fluid in that they shape how we are becoming in ever-changing and mysterious ways. Service-learning can be a significant enhancement to self-understanding, particularly with regard to class, especially if we learn to look for the paradoxes and dialectical relationship between various facets of ourselves.

REFERENCES

Amy (2004). Interview. Central Pennsylvania.

Aronowitz, S. (2003). *How class works.* New Haven, CT: Yale University Press.

Blau, Peter & Otis Duncan (1967). *The American occupational structure.* New York: Wiley.

Boyle-Baise, Marilynne & James, Kilbane (2000). What really happens? A look inside service-learning for multicultural teacher education. *Michigan Journal of Community Service Learning*, 7: 54–64.

Bureau, US Census (2004). *Three-year-average median household income by state: 2000–2002.* US Census Bureau 2002 [cited June 7, 2004].

Bureau, US Census (2004). *Press briefing on 2002 income and poverty estimates.* US Census Bureau 2003 [cited June 7, 2004].

Coleman, Richard & Lee Rainwater (1978). *Social standing in the United States: New dimensions of class.* New York: Basic Books.

College, John Tyler Community (1995). John Tyler Community College Weekend College: The first semester.

Duberman, Lucille (1976). *Social inequality: Class and caste in America.* Philadelphia: Lippincott.

Flower, L. (2002). Intercultural inquiry and the transformation of service. *College English*, 65(2): 181–201.

Green, Ann. (2001). But you aren't white: Racial perceptions and service-learning. *Michigan Journal of Community Service Learning*, Fall: 18–26.

Horvat, Erin & Anthony Antonio (1999). "Hey, those shoes are out of uniform": African American girls in an elite high school and the importance of habitus. *Anthropology and Education Quarterly*, 30(3): 317–342.

Hourigan, Maureen (1998). Critical literacy in "White Ethnic" classrooms: When empowered writing reveals class. Paper read at College Composition and Communication, at Chicago, IL.

Jackman, M. & Jackman R. (1983). *Class awareness in the United States.* Berkeley, CA: University of California Press.

Jacqueline (2004). Interview. Central Pennsylvania.

Jones, Susan. R. & Elisa S. Abes (2004). Enduring influences of service-learning on college students' identity development. *Journal of College Student Development*, 45(2): 149–167.

Katelyn (2004). Interview. Central Pennsylvania.

Kingston, Paul (2000). *The classless society.* Stanford, CA: Stanford University Press.

Lipset, Seymour and Reinhard Bendix (1959). *Social mobility in industrial society.* Berkeley, CA: University of California Press.

McIntosh, Peggy (1988). White privilege and male privilege: A personal account of coming to see correspondences through work in women's studies. Wellesley, MA: Wellesley College Center for Research on Women.

Nieto, Sonia (1998). On becoming American: An exploratory essay. In *A light in dark times: Maxine Greene and the unfinished conversation*, ed. William C. Ayers & Joanne L. Miller. New York: Teachers College Press.

Novek, Eleanor (2000). Tourists in the land of service-learning: Helping middle-class students move from curiousity to commitment. Paper read at National Communication Association, at Seattle, WA.

Ryan, Jack & Charles Sackrey (1996). *Strangers in paradise.* New York: University Press of America.

Terenzini, Patrick T. et al. (1996). First generation college students: Characteristics, experience, and cognitive development. *Research in Higher Education*, 37(1): 1–22.

Zweig, Michael (2000). *The working class majority: America's best kept secret.* Ithaca, NY: Cornell University Press.

4

BEYOND A WORLD OF BINARIES: MY VIEWS ON SERVICE-LEARNING

Tiffany Dacheux

As I read Sue Ellen Henry's chapter on service-learning and the first generation college student, it became even more apparent to me that to be working class at a college with a more or less affluent student population is quite a peculiarity. On the surface, I, the working-class student, look just like any other student, save perhaps for a less splendid wardrobe, car, and various other accoutrements. It is only when one speaks to or interacts with a student like myself that the difference becomes discernable. I lack that casual assuredness that comes from *knowing* that this (meaning college) is where I belong. I tread as though it is borrowed ground, and resort to a self-effacing tone when I must (and indeed it is only then that I dare) ask for any form of assistance.

Just as it did for the women in Ms. Henry's study, college struck me as an unequivocally weird place. Here was this place where well-to-do parents send their children for four (or more) years so that they can come out exactly like themselves for the "bargain" price of approximately $30,000 a year. And, as if that were not enough, the children *expect* their parents to do this, and to provide them with spending money to boot. This atmosphere of entitlement astounded me! I saw students who (in my admittedly biased opinion) had everything claim that they had nothing, that life was not fair, or that, despite not having a job or any such responsibilities, they simply had not had enough time to complete their assignments. As I juggled jobs, family life, and academics, it seemed to me that these students truly did not understand how lucky they were. Having been raised in a city that is arguably on the decline, it struck me as odd that so many could take so much for granted and still want more.

When I began my studies, something else about the college environment struck me. Ever since I had been old enough to want an education, I had sought to get away from the dilapidation of my local school district. I just *knew* that there had to be someplace better. Ever since I had begun to excel in school, I had sensed that there was a gap between my family and me. I could see that I was considered "weird" at best and snobby at worst because I would do things like proofreading my parents' writing or correcting them if they spoke incorrectly. On one level it was appreciated (I was often the one in charge of family correspondence), but in another way it was resented (no adult likes for a 10-year-old to correct them). My family was proud that I did so well in school but I think it was also a cause for concern. What role would they play as I increasingly surpassed them intellectually? It seems that on some level they knew where I was headed: toward identifying things as educated/good or uneducated/bad, a binary that would ultimately call on me to assimilate to one side or the other. Little did they realize that the process was already underway. As any child begins school, she is taught that learning is good and that those who choose not to continue with their education are somehow "wrong." They are encouraged to work hard(er), graduate, and to seek out yet more learning. This presents no clear disjuncture in the lives of middle-class or otherwise well-to-do students, but for working-class children whose parents in many cases never graduated high school, the student may have a difficult judgment to make. By accepting what the school says, the student is effectively condemning her parents' life choices, if not them personally. The visage is gone: her parents are no longer the idyllic beings she had supposed them to be and are instead replaced by what she is taught to *not* want to be: high school dropouts and wage laborers.

So this is the mentality with which I approached education. I wanted something better because I had, in essence, condemned what I had grown up in. Underachievement was certainly not valued and thus I began to dissociate myself from it, at least when with more educated people. At home things were one way, but at school things were quite different. With my family, I would never discuss my aspirations for college or a desire to become a writer but at school this was appreciated and encouraged. When I matriculated to college, this phenomenon became even more acute. My family understood little or nothing about my life at college and those I interacted with at college knew little about my working-class heritage. Thus, I began to develop a set of "parallel" identities, neither of which was complete or entirely true. With family and other people I associated with my working-class beginnings, I was careful to be like them—complaining about the toils of work, using the appropriate colloquial speech, and avoiding coming across as stuck-up. At school and with more educated people, I felt freer: I did not have to stop and make sure I was using appropriate vocabulary or that the context of what I wanted to discuss had been established—I was able to let my conversation flow. But on the other hand, this really was not me either—I had to be careful on the opposite end, in the sense that I did not want to appear uneducated. So in the end, I had to be vigilant in not allowing one identity to spill into the other.

It must be noted that I was not quite aware of this separation in my identity. I thought that I was blending my past with my future, and doing a good job at it too. But what happens when someone like myself or the women in Henry's chapter become aware of the parallelism in their lives? In my case,

I began to become aware of it during my studies in education. As I would read about the "problem" schools, I would sometimes see my own schools and my own experiences. As my classmates shared their thoughts and feelings, I would be struck by how shocked or how disinterested they were.

The real death knell to my ignorance of the parallelism came when I began observing in secondary schools and service-learning in my college community. For my observation, I spent some time at the middle school that I had attended. In the service-learning, I spent time tutoring at an after school center that primarily serves Hispanic children. I assumed that I would be very comfortable in both of these situations—in the first because it was my old school and in the second because the schools I had attended were always very diverse. I thought I could walk in there and talk to the kids in their vernacular (thus switching off the educated tone) and they would accept me since I was like them in so many ways. In a sense, I guess I thought that since I was of both "worlds" I would have an inside track on how to help them. When I actually got to the locations, I was anything but comfortable. I simply could not blend in as seamlessly as I had expected. It was as if a wall was between us. As I thought about the situation, I became aware of how different I had become. I was not entirely like these students nor was I entirely unlike them. It was true that at least emotionally, I could relate to these children's experiences, but I could never again fully share their experiences. Likewise, I was not—and likely will never be—fully assimilated to middle-class culture.

Then I reflected on how I felt when I had been the subject in service-learning projects. While in high school, I had participated in a program designed to help high-achieving, low-income students make it to college. I remembered hating the way that the college student "servers" tried to ana-lyze us and figure out what was wrong with us so they could "fix" it. It was then that I realized what had prevented me from connecting with the students in my own service-learning project: somewhere along the way I had bought into the notion that these kids (poor, brown, and/or urban youths) were somehow "wrong" and needed me as the now educated college student to raise them up. I had placed them and myself into a binary of "served" and "server" and by doing so assumed them to be somehow lacking in comparison. There was a distance between myself and these kids I so desperately wanted to help and that distance was created by the very fact that I thought they needed *help*. In order to develop my own new sense of identity as a college student I had formed this binary, a binary driven by a sense of superiority. It was as if I could only say who I was if I could also say who I was better than.

This is arguably a flaw of service-learning: it encourages the separation of people into a binary and assumes that one side needs to be lifted up by a more capable other. The needy person must acknowledge the wrongness of his previous ways and acquiesce to this higher good. As I thought on this, I began asking myself if what I had—an unconsolidated half-and-half identity—was really a higher good and what the process of service-learning really had to offer these so-called served people. As we place children in service-learning situations, are we not simultaneously telling them that they are unable to get education on their own? Is this not the same thing as telling them they are somehow lacking? Taken as a whole, this seems anathematic to the intended purpose of service-learning. When a student enters in a service-learning situation, the first thing she often learns is of the "plight" of those she will help: their poorness, darkness, or less than satisfactory academic achievements. Difference is emphasized, for the intended purpose of "preparing" the serving students for what they might encounter. What this really does is set them up to "judge" those they might help and place them in the same binary that I did. For students like myself, this holds true as well. Though I felt I could identify with these kids and serve them, I was too hung up on the notion that I was helping them. "Help" it seems is a double-edged sword—we want to help others, but only if the process can make us feel better about ourselves. Thus, a binary of superiority is often formed, even in cases where the server is almost identical to the served.

So, is the intended experience of service-learning lost on people like the women in Henry's study and myself? I do not think so. While I may not have had my eyes opened to the plight of some group of people unknown to me (since I was, in a sense, them already), I was, like those in Henry's study, made aware of the certain degree of privilege I possess. But I think for people like myself the main benefit is something that was not intended: the realization of our peculiar situations as first generation working-class college students and the binaries that may exist in our identities. It is from this realization that we can come to appreciate our modicum of privilege, and work to make it accessible to others. And, only by acknowledging the parallelism in our lives, can we begin to attempt to synthesize it and develop more of a dialectical relationship between our working-class beginnings and educated aspirations.

5

CHANGING PLACES: THEORIZING SPACE AND POWER DYNAMICS IN SERVICE-LEARNING

Caroline Clark and Morris Young

This chapter reflects our long history as collaborators around issues of service-learning and literacy. That history, beginning as co-teachers of an undergraduate service-learning course at the University of Michigan in 1995, has since moved with us across time and institutional contexts into new spaces, places, and understandings of service-learning. Our interest, in this chapter, takes up these issues of time, space, and place, and considers how our theorizing of service-learning has broadened and changed over time. In it, we discuss predominate ways of conceptualizing and theorizing service in undergraduate contexts; our work across service-learning courses and sites; and new possibilities for understanding service-learning in relation to space.

Our past and more recent histories around service have all occurred in the context of higher education, and all have involved literacy learning as a focal point. At the University of Michigan, we co-taught the service-learning course, "Learning Communities." Situated at the university and in two community sites—an after school tutoring program at a local middle school, and a similar program at a neighborhood community center—we tutored, along with our undergraduate students, and engaged in a weekly literacy seminar.[1] More recently, Morris has engaged in initiating a new service-learning program at Miami University. While the Miami service-learning course was developed out of the experience of "Learning Communities," the primary difference from Michigan was that the selected service site at Miami was not a preexisting tutoring program but rather an underserved community that existed on the margins of the larger university community.

Our data for this chapter come from these service-learning sites and courses at Michigan and Miami. Across the years, we have copied and collected the course-related writing from students, including their tutor logs, reading notes, and class papers and assignments. Our attention to students' writing is quite purposeful; we believe that these data provide our best window into understanding how students experience, negotiate, and make meaning from their service work. We draw on these texts, as well as field notes and observations in the sites, in order to illustrate how paying attention to space in new ways may allow us to productively complicate our understandings of service-learning. As we view these texts and experiences through the lens of spatial theory what is revealed are distinct features that provide insights into service-learning in higher education more generally. By foregrounding the spatial dimensions of service-learning, we make a conceptual shift from "changing individuals" and the individual transformation of students to "changing places," and the work that occurs in that shift, including the transformation of students in a broader context of community, culture, and the social dynamics of service.

CONCEPTIONS OF SERVICE-LEARNING, PAST AND PRESENT

Across the 1990s and into the new millennium, service-learning has proliferated on college and university campuses. In 1997, Adler-Kassner, Crooks, and Watters described "a microrevolution in college-level Composition through service-learning" (p. 1). Receiving bipartisan support across two presidential administrations (G. H. W. Bush and Clinton), and support and endorsement across academic disciplines (see, e.g., Kraft and Swadener 1994), service-learning has been hailed by faculty and students for producing "radical transformations of their experiences and understandings of education and its relation to communities outside the campus" (Adler-Kassner, Crooks, and Watters 1997, p. 1). While these effects are certainly laudable, we think they may provide a limited view of the complexities of service-learning.

In part, this is a structural feature of service-learning. While the goal is to link the two, all too frequently, both in higher education and in K-12 contexts, "service" is taken up apart from any "learning" or reflection. Devoid of any reflective activities that require students to connect their experiences in service sites to their classroom work, service-learning becomes, at best, volunteerism and at worst, merely the logging of hours toward a community service graduation requirement. Moreover, such approaches may lead to "logics" of service (Flower 1997) that simply reify positioning "others" as helpless or needy in contrast to their own constructions of self (Clark 2002/2003).

In part, this limited view of service-learning is also conceptual. Service-learning has roots in both notions of *noblesse oblige* (e.g., Addams 1910) and experiential education (e.g., Dewey 1938). These roots continue to shape how service-learning is viewed and understood by students if not instructors. When viewed as simply helping those "less fortunate," students may fail to see the role that their own privilege plays in the dynamics of power (see, e.g., McIntosh 1989). Likewise, when seen as strictly experiential, the focus remains primarily on students, themselves, and what they "get" out of a service-learning experience (Kendall 1990). Service-learning, too, has been framed by theories that attend to issues of civic participation (e.g., Barber 1992). Again, however, while the upshot of such work is service to the community, the central focus is on the students themselves and on producing and reproducing citizens.

More recently, approaches to service-learning have been informed by critical theory and pedagogy (e.g., Freire 1970; McLaren 2003). Here the focus is on issues of power in teaching and learning, including service-learning. Applying such theories to service-learning work bodes to focus students' attention toward activism and social justice. However, while such conceptions attend to reciprocity, mutual learning, and shared benefits across student participants and community members, the ultimate focus is still on transforming the perspectives and worldviews of students.

Often these conceptions of service involve encountering a new place. In higher education, for example, the expectation is that college and university students will get to know the cities around them, break through the "town and gown" divide (Flower 1997) and "rearticulate the college or university as part of rather than opposed to the local community" (Adler-Kassner, Crooks, and Watters 1997, p. 4). By changing places, university and college students will become embedded and invested in community problems, seeing themselves as "intercultural collaborators" rather than simply "commentators on diversity" (Peck, Flower, and Higgins 1995, p. 214). Like many conceptions of service-learning in general, this view of changing places is quite limited. Changing places by leaving their campus and working in a homeless shelter should result in a new perspective on issues for the participating students. Without deep, authentic engagement however, standing in this new place could result in nothing more than a " 'field trip' to *look at* the poor" (Jones, Gilbride-Brown, and Gasiorski, this volume).

Simply moving bodies from one place to another, and standing in someone else's shoes, so to speak, may not do justice to the complexity and significant work of service-learning. As our title suggests, we want to expand and complicate understandings of "changing places" in service-learning by viewing our own work through the lens of spatial theory. We believe that considering our past practices with an eye toward issues of space has revealed important issues we may have overlooked on first glance—new understandings of the micro-politics and day-to-day work of service-learning. Moreover, we believe that spatial theory may also expand general understandings of service-learning, writ large, assisting others as they plan and teach their own service-learning courses.

THEORIZING SERVICE-LEARNING

Our work, both past and present, points to the necessity of theory in service-learning. In many ways, the theorizing we have done around our own service-learning work has been limited. For example, the theoretical lenses employed at the Michigan site focused on developing critical reflection and an understanding of larger contexts for literacy, community, and service. Throughout various iterations of the course, instructors drew on theory in teacher research (Cochran-Smith and Lytle 1993), narrative theory (Bruner 1986, 1994), and critical theory and pedagogy (Freire 1970; hooks 1994) to frame their instruction. As described earlier, undergraduate students engaged in tutoring twice a week at two community-based sites while meeting, once a week, for a literacy seminar on campus. Our overall philosophy was aimed at enabling students to reflect productively on the nature of literacy and their service experiences through the acts of reading, writing, and dialogue.

While the Miami course incorporated many of the strategies and practices of the Michigan course, the Miami course also required a type of theorizing and attention to developing a conceptual model of a community literacy program since it was being created specifically out of the work of this class. In this instance, the course members worked with the community members to understand the literacy needs and practices of the community. Students surveyed the community members to find out what kinds of activities and services they would value, from book clubs and writing groups for adults, to conversational English practice, to an after-school program for the community children. The result of the survey was the creation of an after-school program for children that met twice a week.

In addition to bringing theory to bear in our delivery of these courses, we continually theorized "after"—looking back at our teaching, our students' writings, the experiences at our services sites, and so on—in order to refine our approaches to teaching and to analyze our understandings of how service-learning "works." Examples of this at Michigan include addressing the connections to literacy (Nye and Young 1999), bringing literary criticism (e.g., Bakhtin 1981) to bear (Minter, Gere, and Keller-Cohen 1995), and using "positioning" theory (e.g., Davies and Harré 1990) to better understand service relationships (Clark 2002/2003). In all of these accounts, however, theory concealed as much as it revealed. As Eisner (1993) points out,

> The theoretical constructs with which we work define in large measure the features of the universe we are likely to see. The visions that we secure from the theoretical portholes through which we peer also obscure those aspects of the territory they foreclose. And foreclose they do. (p. viii)

The application of positioning theory to service-learning provides a good example of this foreclosure. In that work (Clark 2002/2003), the focus was on individual service relationships between tutors and the children with whom they worked. The focus was on tensions between individual agency as revealed through interactions and jointly produced storylines (Davies and Harré 1990). Like much service-learning work, the locus of theory was the individual—the "radical transformations" (Adler-Kassner, Crooks, and Watters 1997) of the undergraduate students engaged in service. As such, positioning theory left out the broader dimensions of institutions, contexts, power, and time—essentially, "space" as it is understood in spatial theory.

SPACE / PLACE AND POWER

Our understandings of space/place and power in service-learning are informed by recent work focused on "reasserting space" into social analysis (Wilson 2000). Here space/place is not simply a neutral, physical location or setting where activity takes place. Rather, "recent theorizing about space has brought to light that space is a product and process of socially dynamic relations" (Sheehy and Leader 2004, p. 1). As such, space itself is "a fundamental constituent of knowledge/power regimes" (Soja 1989; Wilson 2000). Taking up space as dynamic and always changing rather than stable and fixed raises questions of how spaces are changed, when this occurs, and what these "new" spaces become (Harvey 1996, in Sheehy and Leander 2004, p. 2). Circulating through these relations are issues of power.

Conceptualizing space as both product and process entails seeing space as more than simply the backdrop against which social practice occurs; rather, "lived space" (Lefebvre 1991) occurs in the trialectal relation among spatial practice, representations of space, and representational space. Leander (2004) refers to these as "perceived, conceived, and lived forms of space" (p. 118). Perceived space involves "production and reproduction of relationships between people and things, and people and practice" (Sheehy 2004, p. 95). For example, service work that involves "tutoring" assumes a set of practices and relationships among participants—assistance with some academic task, a tutor or more knowledgeable person assisting a tutee or less knowledgeable person, and so forth. Conceived space involves people's abstract understandings of space and the role of knowledge/power in shaping how people engage in perceived space. As Sheehy argues, "Power functions as people relate around objects. That one object of study exists and not another is itself indicative of power relations" (Sheehy 2004, p. 96). Taking the service as "tutoring" example, again, one can imagine work around academic texts such as worksheets, reading textbook chapters, and producing writing related to homework as acceptable objects of study. These objects carry with them the power and authority of schooling. While other objects of interests may exist in a "tutoring" space—for example, playing games, making crafts, or engaging in sports—these must compete and contend in power relationships with the official purposes of "tutoring." Finally, lived space involves the bodily experience of space as well as the space of imagination. In "tutoring" service, this might entail how bodies are engaged in tutoring—generally sitting side by side, at a desk or table, sitting still and focusing on some text of study; as well as the imagined functions or purposes for the tutoring tasks—completing homework or studying in order to improve grade and performance in school. All of these spaces—perceived, conceived, and lived—function simultaneously. With these theoretical conceptions of space in mind, we take up the question of "changing places" in service-learning.

SPATIALIZING SERVICE-LEARNING

Spatial theory turns our attention from "changing individuals" and the individual transformation of students to "changing places," and the work that occurs in that shift. This shift invites new questions as we reflect on old and plan for new service-learning ventures. Echoing Wilson's (2000) questions, we're inclined to ask, by what social processes is space constructed? How does place shape/produce power in service-learning? We contend that part of the work of service-learning is "the making of a new space" and "the flow of information exchanged" in such spaces (Sheehy 2004). In her work examining curriculum change in a 7th grade civics class, Sheehy describes what she calls "thick places"—places where flows of information tend to be patterned and stable (e.g., the IRE sequence in classroom discourse).[2] Here there is little student-to-student interaction, and objects of study do not relate to life outside the classroom except in terms of invisible, ideological force (e.g., information presented in class will be required for some later assessment). These thick places as Sheehy (2004) describes them, tend to reproduce themselves. In contrast, she describes "thin places"—a new space premised on different flows of information, patterns of exchange, and bodily experiences.

> Unlike exchanges in the thick place, students in the thin place . . . are not focusing on the object in order to reproduce it in the invisible power relations in which it was offered to them. Rather, they are claiming their own object, expressly to situate themselves in specific power relations. They do not attempt to replicate an object at all; they are negotiating its shape, taking the object from a number of relations in which it was first produced in the classroom and placing it within yet other ones.   (p. 100)

As places change, thick places may become thin places, and in the interim comes what Sheehy (2004) calls "in-between space"—a frenzied, unpredictable space where the tensions around change processes are revealed.

In our own experiences, and in our readings of other service-learning work, we see much evidence of service sites as thick places—places of patterned and stable flows of information and exchange. Whether these are homeless shelters, meal-delivery programs serving individuals with AIDS, or school-based tutoring programs, they are fairly stable. And, while undergraduates working within these sites may be changed through their experiences, the expectation is that volunteers will assist in the site as is. As such, these sites tend to reproduce themselves and socialize the volunteers to fit within them rather than the other way around (much as school functions in the lives of students). At times, however, students may enter these sites as resistant (see, e.g., Jones, Gilbride-Brown, and Gasiorski, this volume). Or, students may carry constructions of objects of exchange, practices, or flows of information that challenge established notions within the sites. In such cases, these sites may be better understood as in-between places. In each place—thick, thin, and in-between—the trialectics of space play out.

The Michigan site, for example, seemed illustrative of a thick place. The service sites had on-site staff, and histories in the community that predated the university service-learning course. Our approach might be best understood as constructivist. Rather than assign materials, texts, or activities to be accomplished in our twice-weekly work at the sites, we allowed tutors and the children with whom they worked to figure this out together based on the children's needs and interests. The service sites maintained materials (papers, pencils, supplies, games) to support these general tasks. In addition, children were expected by the service site staff to bring school-based work as well. The sites housed desks, tables, chairs, partitions, and in the Southside case, one private room where tutors and children could work. As a thick space, then, the service sites in Michigan presupposed certain social practices, tied to power and knowledge relations in the site and beyond (including school), and invited particular bodily experiences of the space. Our course, in turn, focused on literacy. And while we discussed functional constructions of literacy as reading and writing, through readings and class discussions, we intentionally pushed students to broaden their conceptions of literacy and to consider the social, cultural, critical, and political aspects of literacy. Bringing these notions of literacy into the service sites created tensions in the space as it changed from a thick place to, at least, an in-between place, if not a new, thin place.

On first examination, we might expect the after-school program at Miami to be an alternative to the thick place that we see at Michigan since students entered into the site looking to create activities and practices based on the needs and desires of the children who attended. That is, while the program on one level functioned to assist children with homework, on another level the program provided opportunities to reimagine literacy practices outside of a classroom context and within the home community of the children, looking to draw on the talents and interests of both the children and "tutors." However, despite the opportunity to create something "new," the after-school program still functioned as a thick place on a number of levels. First, its location at The Manor, the university residence complex for graduate students and students with families, placed it at the physical boundaries of the campus in buildings originally built in the late 1950s. Second, unlike the undergraduate residence halls designed in the modified Georgian style that characterizes the university campus, the Manor resembled mass-produced housing projects that are more often associated with low-income neighborhoods, reinforced by the presence of clotheslines hanging outside often filled with laundry out to dry, despite the middle-class and professional-class backgrounds of most of these families. Third, because many international families chose to live there, it acquired a representation as an ethnic enclave, reinforcing another stereotype of "foreigners" who cluster among themselves and choose to isolate themselves from mainstream American culture; the Manor even picked up a derogatory moniker as the "Asian Invasion." Thus while the Manor was associated with the university, specific stereotypes generated around race, ethnicity, and social class marked this site because on the surface it did not fit so easily into a university culture focused primarily on undergraduates.

The after-school program itself initially functioned as a thick place by virtue of the work that the university students imagined they would do there: assisting with homework, working with nonnative speakers of English on their English language skills, and literacy enrichment through the accepted practices of reading and writing. However, the community also saw the after-school program as an educational opportunity for their children, resulting in steady participation by the children and continued support from the parents. As the program progressed and we learned more about this community and the range of literate activity by the children, the students and the children began to challenge the boundaries of the space. In one sense, the Manor community's own understanding of itself constituted another thick place, but one that began to disrupt others' construction of it. As one student, Sherry, wrote in her journal:

> the idea of community is mostly just that—an idea, created through rhetorical means. Stated another way: in order for a community to exist, there must be a certain degree of awareness among its members. Either they must recognize the values that they share, or (in the case of a geographic community) they must identify the factors that group them with one another and apart from other people.
>
> I see Miami Manor as a community in this complex sense of the word. There is a wide variation in the backgrounds of the residents, yet they all find themselves in similar living situations right now [. . .] I'm reminded of what [the resident assistant] told us when he first came to speak to our class. He mentioned that the residents can sometimes feel alienated, being mostly "nontraditional" or foreign students and living in housing that's not quite a residence hall and not quite an apartment building. This may be the most basic bond that they share: the feeling of not fitting in to the neatly defined categories of a small college town.   (STL 2/11)

What happens, then, when competing understandings of a place and its social practices exist within the same space? How may this result in a reimagining of the place by a changing community membership? What happens when a sense of place becomes unstable in the face of different pressures from a larger context?

UNDERSTANDING SERVICE AS CHANGING PLACE

Across the Michigan and Miami sites a tension emerged around the purpose of service—specifically tutoring. In both sites, for example, the undergraduate students struggled with their perceptions of why children came to the sites. Their sense of perceived space/spatial practice, as well as the representations of space/conceived space and the objects around which tutoring occurred were often in flux. Likewise, as we see, the lived space/bodily space that was physically experienced was troubled as place was changed.

Most of us from the universities—students and instructors alike—came from middle-class homes, and most had experienced success in schools. Entering the tutoring situation from this perspective caused the university students to expect the "problem" of school-success to reside in the child with whom they were working. Furthermore, university tutors entered expecting the children to "need" them. Clearly, if these youngsters were attending after-school tutoring sessions they needed academic help. In many ways, then, the university students' expectations mapped on well to the institutional norms of the site, and to the material exchange of ideas, information, and objects in place within these thick spaces—both real and imagined. As academically successful themselves, they enjoyed competence with conventional literacy learning tasks, and they were prepared to share these with the children, thus reproducing the space at hand. Tensions arose, however, in at least two areas. In part, the new constructions of literacy that the university students carried into these spaces challenged the existing objects around which tutoring occurred. Likewise, the children at the sites exerted their own agency in changing place.

As part of the social network at play in the space, emerging understandings of literacy brought from the university service-learning seminars, disrupted the established sense of what counted as tutoring in the site. For example, Seth, one of the Michigan university tutors, consistently constructs home-work as the appropriate object around which tutoring work should occur. Without it, his sense of the purpose of tutoring and of this space is changed. In his tutor log, he writes:

> Today was a very difficult day. David never comes with work, and he spends the whole time disrupting Kevin. I told him that I was not there to baby-sit, but rather to help and that he needed to be serious about his studies before we can have fun. (STL 2/6)

Without homework, Seth's understanding of place shifts from tutoring to babysitting. Later, when David does come with homework, tensions are still at hand.

> Today David came with work. Unfortunately, he didn't also come with instructions. I tried to figure out how to do his paper with him, but he had trouble explaining and didn't have any patience with letting me help him. He seemed to not think that I knew what I was doing. (STL 2/8)

In this in-between space, as the thin space being created confronted the thick space at hand, tensions around Seth's level of competence in spatial practice emerged (Sheehy 2004). Even with the presence of homework, without adequate instructions, he and David could not engage in the spatial practice of "tutoring." Similar tensions occurred at the Miami site. Kerri, for example, writes:

> This past Wednesday at Miami Manor, I was somewhat disturbed by my role in the program. I worked with Sarah for most of the hour, and I felt sort of

badly about the experience as a whole. Because she is younger than the other kids, and because her language skills are less developed and her attention span is much shorter, I felt more like a babysitter than anything else. . . . I'm sure, at some level, Sarah benefits from the raw contact and English language interaction. But I still feel inadequate and somewhat useless when I follow her back and forth across the room from drawings of snakes to the same *101 Dalmatians* book.   (KTL 2/19)

Here, again, tensions emerged around Kerri's level of competence in spatial practice. As the space changes, Kerri is "disturbed" by her role in the space and questions her adequacy. In both cases, Seth and Kerri are troubled as the social relations of tutoring fail to reproduce themselves in this changed place.

Across the sites, the children continued to change the space as well, exerting their own power. Often, this occurred in terms of lived/bodily space. In the case of Sarah, for example, her movement back and forth, across the room troubled Kerri's sense of the "tutoring" space. Such physical expressions did not fit Kerri's real or imagined sense of tutoring. Likewise, Seth writes repeatedly about his two students, Kevin and David, and his tensions around lived/bodily space.

> I worked hard on getting Kevin to sit still and work. He had to do it in spurts, taking breaks to keep himself occupied. One thing I especially noticed about him is the fact that he seems to be somewhat starved for affection. Whenever I am working with him, he seems to want to always sit on my lap and jump on me. I wonder if it is age appropriate for him.   (STL 1/24)

Later, in the same tutor log entry, he describes both boys as "disruptive" and having "very much difficulty sitting still and working on homework. . . ." Across the weeks, he writes again and again about David "disrupting" the work of tutoring. And, while he expresses satisfaction at Kevin's improvement in terms of spelling, he is still upset by his bodily engagement, writing:

> He [*Kevin*] has become a better speller, but he still can't sit still when doing it. Spelling the words he's always jumping around the room and under the table.   (STL 2/13)

So, even though the objects, around which tutoring occurs, fit Seth's expectations of the space and of literacy, they still do not reproduce the social relations he assumes of tutoring. For both Seth and Kerri, the bodily actions of the children, as well as the objects in circulation (drawings of snakes, a single picture book) do not fit their logics of "tutoring" space.

The most direct transgression of bodily space occurred midway in Seth's work with David. In his tutor log, he recounts the incident:

> Today I went and had my toughest time yet. It was Valentine's Day so none of the kids wanted to work. That was fine. I was willing to give the kids a day to just play with us. But I wasn't ready for what happened. Kevin and I were playing

a game similar to Chinese checkers, and David just didn't want to play. He disrupted our game and ran around messing with everyone. Then he took my hat. I tried to be patient and didn't pay much attention to his antics. Then he ran into the bathroom. I was really worried he was throwing my hat down the toilet. When he came out the hat was nowhere to be found, except when he showed me that he had put it down his pants. Boy, did that piss me off. I had to go outside to cool off or else I really might have gone off on him. He came out. I told him he had no respect for me, for other's property, for what the purpose of tutoring was all about. He looked very upset. He said he was sorry. I told him I didn't know why he even bothered coming to the center when all he did was make trouble. He said he liked to meet people. I told him I wanted him to seriously think about what he wanted out of the center and be prepared on Monday. He agreed. (STL 2/14)

Throughout this description, Seth attends to David's bodily disruption of the tutoring space. Even before placing Seth's hat down his pants, David is implicated for disrupting a game and "messing" with everyone. Seth's response to David's physical experience of the space is to not pay attention to his antics. But for David, his antics are the space. He comes because he likes to meet people; for him, this is the purpose of tutoring. Again, then, we have evidence of an in-between space as Seth's thick space of tutoring meets with David's sense of the space.

When Seth returned to the site, on Monday, he had given some thought to David's reasons for coming to the after-school site, and he came prepared to accommodate David's purposes for tutoring—to meet people. While this move does reflect a change or transformation on the part of Seth, we contend that it also reflects a shift in space as Seth works with David to renegotiate and change place. Seth writes:

As bad as the last time went, this time went good. I came in with the attitude that the center was no longer going to be a place where I could come to teach; rather, I would approach it as a place to get to know my kids and let them establish how much learning would take place. The kids loved it. I joked with them, calling them punks and whatnot. They learned that they could just talk to me, and it was their decision as to how much work they would really accomplish. (STL 2/21)

This shift entails change in terms of the social practices and objects of exchange in the space (e.g., jokes, opportunities to talk, and get to know one another); in the ideologies that will hold sway in this space (e.g., tutoring as not teaching); and the lived experience of the space, as Seth writes about Kevin's bodily engagement,

Kevin had some more spelling words. . . . He gets frustrated and starts jumping around the room. When he does this I try to just wait it out until he decides that he can do it. I know that eventually he will need to learn to sit and concentrate, but for now I think this is the best method. (STL 2/26)

With David, Seth works even further to change place, creating a thin place of spatial practice, conceived space, and bodily space that he and David negotiate together:

> David and I worked alone. I had a great idea to bind us. I created a common enemy. Any time that some kind of foreigner (i.e., another kid) walked into the room, David and I would scream, "Get outta here sucka!" at them until they left. In doing this, David and I were able to become allies. He brought some kind of worksheet, which I helped him with. He is learning to trust me more socially and academically.   (STL 2/28)

In this thin place, Seth and David are clearly not "focusing on the object in order to reproduce it in the invisible power relations in which it was offered to them. Rather, they are claiming their own object, expressly to situate themselves in specific power relations" (Sheehy 2004, p. 100).

Another example of the transformation of the bodily disruption of space into a new social relation can be seen in the transgression of space where we see flow between defined spheres of social relations: the after-school program and the home community. While the location of the Miami after-school program at the Manor placed it on the margins of the larger university community, this liminality more easily allowed for exchanges between "home" and after-school program as well as allowed the students to better understand the social practices of the children in their own environment. This flow between places and the bodily disruption of space not only by the children but also by the tutors is illustrated in the games generated by the community. One student, Lisa, observed in her journal how another student crossed over into the children's space:

> After class today, Sandra and I discussed how she played tag with the kids before and after the program on Monday. She explained that all of the kids played together, which I find interesting and encouraging considering that there is a wide age range and both genders involved. This fact got me thinking about the typical relations between girls and boys at any of the kids' ages. In most situations, there is teasing or at least a division between the two, but in our case, all of the children cooperate and even seem to be friends. For the older boys to readily help a younger child spell a word is impressive. I am curious as to what kind of bond links these children. Is it living in such a physically small community? Is it that they were all born in a country other than the United States, and therefore all feel like foreigners to an extent? I also wonder how we, as older outsiders, fit into their community, which had been formed before merely by living together as neighbors.   (LTL 3/21)

Lisa begins to do important work in her journal as she unpacks the meaning of community and how social relations are structured within that community. In her mind, there is something different at work at the Manor since the children seem to work together in ways, perhaps even care for each other in ways that seem atypical for boys and girls of this age. Lisa also seems intrigued by Sandra's crossing over into the children' space because it seems to give

Sandra access to that community, and perhaps even begins to create a new community and social relations as her body disrupts the children's space. However, unlike the uneasiness that the tutors feel when the children disrupt the tutoring space, the children are not bothered when someone new enters their space.

However, rather than simply an exchange of places, this flow between spaces can inform the "work" that occurs in each space. For example, Lisa sees the creation of games as a way to understand and explore the literate activity of the children. While not trying to co-opt or institutionalize these games, she does try to bring the play/work of one space into another that begins to disrupt the thick space of tutoring:

> Sandra also told me that the tag game they played on Monday was "Color Tag," which the boys had invented. For today's project, I thought it might be successful to have the kids write out the rules to Color Tag as a group and then make up their own versions of tag on their own. It took a little prodding at first, but eventually most of the kids wrote out the rules to at least one kind of tag game. (LTL 3/21)

While the children are resistant to some degree, their willingness to bring their play into the after-school program suggests that they are also beginning to understand and to reimagine that space. On one hand, the writing down of "rules" may act to transform a thin place into a thick place, as social relations and practices become reified, no longer the freewheeling play of a child's game. However, the fact that Lisa encourages the children to transform their games, to make them something new and their own indicates that there is a flexibility and flow of ideas and practices. What verifies this new thin place is Lisa's new willingness to cross a boundary and to reimagine herself and the community space she has become a member of. As she writes in her journal:

> After the program, I played Color Tag with Ava, Zoe, Amy, Cara, and Shane. It was really a lot of fun, and it is exciting to me that they want me to play with them. It indicates that they aren't intimidated by me. Hopefully my willingness to play with them indicates to them that I enjoy spending time with them, as well. Even though it is probably easier to form good relationships between tutor and tutee when the tutee is a child, I am still thankful that I have this opportunity to spend time with the kids outside of the "classroom." The more time I spend with them, the more I will learn about their needs, desires, and interests as people, and not just as non-native English speakers. (LTL 3/21)

Lisa has begun to understand her work as a tutor in different terms. While the kids have welcomed her and the others into their community, Lisa needed to be willing to step outside of what she perceived her role to be, to be willing to "learn" about the children and to not simply tutor them.

What Lisa experienced was not unique among the students. While perhaps not all reimagining is fully a new space where we see the free flow of ideas and practices, the students in Learning Communities at Miami did undergo a transformation that helped them understand literacy and service in more complicated ways. As another student, Brenda, wrote in her journal:

> Maybe I am too concerned with getting a tangible "product" from our experience. The [Jeff Grabill] article warns against literacy programs that simply hand out products for learners to consume. I suppose it could work the other way too? That the ultimate goal of a program should not be to simply accumulate a body of articles produced by learners, but to give them a meaningful, enriching experience. I guess I should just accept that on some days, the kids may not be interested in producing anything that can be directly associated with literacy. I should look at the overall picture of how our program can benefit the kids in the long run, and not worry so much about the successes and failures on individual days.   (BTL 2/14)

Brenda's emerging awareness of what a "meaningful, enriching experience" can be can also be read in spatial terms. Rather than be invested solely in the transformation of individual lives or in specific successes and failures, understanding how space is dynamic and will change meanings and social relations as individuals move in and out, as communities ebb and flow, will help transform the meaning of service-learning.

IMPLICATIONS

Applying the lens of spatial theory to our own service work informs our understanding of past practices, our plans for future service-learning work, and the field of service-learning in general. Looking closely at the day-to-day work of service-learning in the Michigan and Miami contexts reveals both why and how to look at service differently—to view service-learning as "changing places." Why look at service-learning differently? Because as our discussion of predominate approaches to service-learning shows, despite the application of new approaches and new theories over time, much service-learning work in higher education persists in focusing on the transformation of individual student participants. While this focus on transforming students may attend to issues of power and politics by examining issues of privilege and attending to issues of reciprocity, often it does not. And, even when the focus is on the "done to" as well as the "doer" (Radest 1993), analyses that treat power dynamics as individual negotiations (between tutor and child, for example) come up short. When confronted with the politics of these negotiations, students are often frustrated and angered by the other (as Seth was, initially, with David and Kevin); or feel guilty, attacked, and hurt when forced to acknowledge their own power and privilege in relation to service-learning work (see Jones, Gilbride-Brown, and Gasiorski, this volume). By holding themselves or their service partners/recipients individually responsible for broader social ills, students may end up feeling alienated from the people they most want to "help" or put off from any future service at all.

As for how to look at service-learning differently, we suggest paying attention to space/place. In short, space matters. As Soja asserts, "we must be insistently aware of how space can be made to hide consequences from us, how relations of power and disciplines are inscribed into the apparently innocent spatiality of social life, how human geographies are filled with politics and power" (1989, p. 6). Looking at service-learning in terms of space/place allows us to see service-learning as work that goes beyond the transformation of individual students through service experiences. Rather, it involves the complex work of changing place. This, in turn, entails changes in all aspects of "space"—perceived, conceived, and bodily experience. As place is changed—from thick space, to in-between space, to thin space—tensions emerge. Understanding these tensions may help us better understand instances of resistance in service-learning. They may also enable us to rethink ways of teaching our service-learning courses. Knowing that the work of changing place entails shifts in social practices that seem logical or natural and are, generally, lived out as such by the many individuals in space reveals the scope of service-learning work. In our case, university students were engaged in "tutoring" work around literacy with young people in communities. In many ways, they came prepared to reproduce the "thick places" they expected these spaces to be. They brought expectations about tutoring spaces—its social practices, the ideologies that underlie it, and its bodily expressions. Likewise, they brought expectations and assumptions about the surrounding community spaces, and the young people and literacy practices to be found there—or not. Our courses worked to introduce them to new conceptions of literacy, as well as new perspectives on the communities where they served. But, as Sheehy (2004) found in her own work on changing classroom practice,

> . . . inserting new objects and new practices into school could not simply replace old practices. Students themselves required old space to make new space. They could not begin a new practice absent of past practice. They had to use the familiar to build the new. Yet, in university classrooms, we often do not speak about space. We treat objects as though they were neutral, as though any object can be inserted into any space. (pp. 111–112)

Sheehy's words remind us of the need to make the work of service explicit. Moreover, they show us that the explicitness needs to occur in articulation with issues of space. Service sites, alone, are not educative, nor are experiences within them. Students have a tendency to see only perceived space— the social practices and daily routines. What they don't see—can't often see—are the conceived dimensions of space; the ideologies that make certain practices seem "natural" or "right." Often, it is through lived, bodily experiences of social life that these assumptions of space are disrupted or contradicted—as was seen in the work of "tutoring" at both the Michigan and Miami sites. Paying attention to all of these dimensions of space should be part of the work of service-learning courses.

NOTES

1. For detailed descriptions of this course, these sites, and our approaches see Clark 2002/2003; Nye & Young 1999; Schutz & Gere 1998; Gere & Sinor 1997; Minter & Schweingruber 1996; Sanchez et al. 1996; Tracey et al. 1996; Minter, Gere, & Keller-Cohen 1995.
2. The IRE sequence is a form of classroom exchange involving teacher initiative, and student response, followed by teacher evaluation (Cazden 1988).

REFERENCES

Adler-Kassner, L., Crooks, R., & Watters, A. (1997). Service learning and composition at the crossroads (pp. 1–18). In L. Adler-Kassner, R. Crooks, & A. Watters (Eds.), *Writing the community: Concepts and models for service-learning in composition.* Washington, DC: American Association for Higher Education.

Addams, J. (1910). *Twenty years at Hull-house.* New York: The Macmillan Company.

Bakhtin, M. M. (1981). *The dialogic imagination: Four essays.* Austin: University of Texas Press.

Barber, B. R. (1992). *An aristocracy of everyone: The politics of education and the future of America.* New York: Ballantine Books.

Bruner, J. (1986). *Actual minds, possible worlds.* Cambridge, MA: Harvard University Press.

Bruner, J. (1994). Life as narrative. In A. H. Dyson & C. Genishi (Eds.), *The need for story: Cultural diversity in classroom and community* (pp. 28–37). Urbana, IL: National Council of Teachers of English.

Cazden, Courtney B. 1988. *Classroom discourse: The language of teaching and learning.* Portsmouth, NH: Heinemann.

Clark, C. (2002/2003). Unfolding narratives of service learning: Reflections on teaching, literacy, and positioning in service relationships. *Journal of Adolescent and Adult Literacy,* 46(4): 288–297.

Cochran-Smith, M. & Lytle, S. L. (1993). *Inside/outside: Teacher research and knowledge.* New York: Teachers College Press.

Davies, B. & Harré, R. (1990). Positioning: The discursive production of selves. *Journal for the Theory of Social Behaviour.* 20(1): 43–63.

Dewey, J. (1938). *Experience and education.* New York: Macmillan.

Eisner, D. W. (1993). Foreword. In D. J. Fliners & G. E. Mills (Eds.), *Theory and concepts in qualitative research: Perspectives from the field* (pp. vii–ix). New York: Teachers College Press.

Flower, L. (1997). Partners in inquiry: A logic for community outreach (pp. 95–118). In L. Adler-Kassner, R. Crooks, & A. Watters (Eds.), *Writing the community: Concepts and models for service-learning in composition.*

Freire, P. (1970). *Pedagogy of the oppressed.* New York: Continuum.

Gere, A. R. & Sinor, J. (1997). Composing service learning. *The Writing Instructor* 16 (Winter): 53–64.

Grabill, J. (2001). *Community literacy programs and the politics of change.* New York: SUNY Press.

Herzberg, B. (1994). Community service and critical teaching. *College Composition and Communication,* 45(3): 307–319.

hooks, b. (1994). *Teaching to transgress: Education as the practice of freedom.* New York: Routledge.

Jones, S. R., Gilbride-Brown, J., & Gasiorski, A. (2005). Getting inside the "underside" of service learning: Student resistance and possibilities. In D. W. Butin (Ed.), *Service-learning in higher education: Critical issues and directions*. New York: Palgrave.

Kendall, J. (1990). Principles of good practice in combining service and learning. In *Combining service and learning: A resource book for community and public service Vol. I*, National Society for Internships and Experiential Education, Raleigh, NC: NSIEE.

Kraft, R. J. & Swadener M. (Eds.) (1994). *Building community: Service learning in the academic disciplines*. Denver, CO: Colorado Campus Compact.

Kumashiro, K. K. (2002). Against repetition: Addressing resistance to anti-oppressive change in the practices of learning, teaching, supervising, and researching. *Harvard Educational Review*, 72: 67–92.

Leander, K. (2004). Reading the spatial histories of positioning in a classroom literacy event (pp. 115–142). In K. M. Leander & M. Sheehy (Eds.), *Spatializing literacy research and practice*. New York: Peter Lang.

Lefebvre, H. (1991). *The production of space*. Cambridge, MA: Blackwell.

McIntosh, P. (1989). White privilege: Unpacking the invisible knapsack. *Peace and Freedom*, July/August: 10–12.

McLaren, P. (2003). Critical pedagogy: A look at the major concepts. In A. Darder, M. Baltodano, & R. D. Torres (Eds.), *The critical pedagogy reader* (pp. 69–96). New York, NY: RoutledgeFalmer.

Minter, D. W., Gere, A. R., & Keller-Cohen, D. (1995). Learning literacies. *College English*, 57: 669–686.

Minter, D. W. & Schweingruber, H. (1996). The instructional challenge of community service learning. *Michigan Journal of Community Service learning* 2 (Fall): 92–102.

Nye, E. & Young, M. (1999). Service learning and the literacy connection (pp. 69–96). In R. A. Sudol & A. S. Horning (Eds.), *The literacy connection*. Cresskill, NJ: Hampton Press, Inc.

Peck, W. C., Flower, L., & Higgins, L. (1995). Community literacy. *College Composition and Communication*, 46(2): 199–222.

Radest, H. B. (1993). *Community service: Encounter with strangers*. Westport, CT: Praeger.

Sanchez, A., Lindsey, T., Jeric, T. J., & Minter, D. W. (1996). Learning to do college. *Michigan Academician*, 23: 431–446.

Schutz, A. & Gere, A. R. (1998). Service learning and English studies: Rethinking "public" service. *College English*, 60 (February): 129–149.

Sheehy, M. (2004). Between a thick and a thin place: Changing literacy practices (pp. 91–114). In K. M. Leander & M. Sheehy (Eds.), *Spatializing literacy research and practice*. New York: Peter Lang.

Sheehy, M. & Leander, K. M. (2004). Introduction (pp. 1–13). In K. M. Leander & M. Sheehy (Eds.) *Spatializing literacy research and practice*. New York: Peter Lang.

Soja, E. (1989). *Postmodern geographies: The reassertion of space in critical social theory*. London: Verso.

Tracey, J., Tingling, F., Goldberg, J., & Gere, A. R. (1996). Why service? Why now? A self-reflective study of service learning's effects on undergraduates. *Michigan Academician*, 23: 447–462.

Wilson, A. L. (2000). Place matters: Producing power and identity. Downloaded from http://www.edst.educ.ubc.ca/aerc/2000/wilsona1-final.PDF.

6

Service-Learning as Postmodern Pedagogy

Dan W. Butin

Introduction

Stanley Fish has done it again. He has thoroughly antagonized liberals with his outrageous statements and at the same time frustrated conservatives intent on appropriating his rhetoric (though conservatives don't usually get the irony; but more on that later).

On his way out as dean of the College of Liberal Arts and Sciences at the University of Illinois at Chicago, Fish fired off a series of essays on the roles and responsibilities of higher education (2003, 2004a). Stick to questions about the truth, he suggested, and don't trouble with issues of morality, democracy, or social justice: "We should look to the practices in our own shop, narrowly conceived, before we set out to alter the entire world by forming moral character, or fashioning democratic citizens, or combating globalization, or embracing globalization, or anything else" (2004a, A23).

The response was swift from both sides. Liberals decried and conservatives embraced Fish's limited notion of higher education. Liberals saw it as an attack undermining higher education's responsibilities to the local and global communities, to participating in the public and democratic sphere, and to making knowledge relevant to and for the polity. Conservatives saw it as a reaffirmation of their claims of the liberal bias in the academy and thus supportive of their demands to get (liberal) politics out of the college classroom.

This chapter uses this seemingly local flare-up as an opening to suggest an alternative narrative for service-learning. Specifically, I want to do the seemingly heretical and embrace Stanley Fish in order to strengthen the theoretical grounding and pragmatic enactment of service-learning as postmodern pedagogy. In so doing, I question the traditional modes by which we speak about and do service-learning. I argue that service-learning as postmodern pedagogy is a direct outgrowth of our "postmodern condition" and to ignore this is to risk compromising the transformative potential of service-learning for the safety of a normative, instrumental, and nonthreatening educational reform model.

This chapter proceeds by explicating four distinct conceptualizations of service-learning—technical, cultural, political, and postmodern—before turning to a more detailed engagement with Fish's arguments. This engagement is meant to show that Fish's critique is valid only for the cultural and political versions of service-learning. In fact, I argue that Fish's stance is not only compatible with but also actually highly supportive of the technical and postmodern versions of service-learning. This chapter thus details the implications of service-learning as postmodern pedagogy and concludes with an overview of the implications for the service-learning field.

FOUR CONCEPTUAL MODELS OF SERVICE-LEARNING

Wittgenstein argued that the limits of our language are the limits of our world (Wittgenstein 2000). The conceptual models we hold determine the world we see (and don't see) and the implications thereof. The way we view service-learning will as such determine what we do with it and how we do it. I propose that service-learning be viewed through four distinct lenses—technical, cultural, political, and postmodern (see Butin 2003a, for a more thorough articulation of this perspective). The value of this typology vis-à-vis other articulations (e.g., Furco 1996; Furco and Billig, 2002; Kendall 1990; Lisman 1998; Liu 1995; MJCSL 2001; Morton 1995; Sigmon 1979) is the lack of a normative or teleological framework (i.e., a standard, "normal," or presumed vision of what service-learning is/should be) and the disentangling of the multiple and usually conflated goals of service-learning.

A technical conceptualization of service-learning is focused on its pedagogical effectiveness. An excellent way, for example, to understand the impact of poverty on families is to work with and observe actual families in actual poverty within the context of an academic course that uses multiple other texts, reflections, and assignments. Likewise with understanding how to teach math to elementary-aged school children or how to articulate and develop strategic marketing and organizational plans for nonprofit organizations. Service-learning is here conceptualized as one among multiple pedagogical strategies; it serves the function of better teaching for better learning. Much like lectures, cooperative learning, or experiential learning activities do in their respective situations.

A cultural conceptualization of service-learning is focused on the meanings of the practice for the individuals and institutions involved. Service-learning is seen as a means to, for example, help students increase their tolerance and respect for diversity, for academic institutions to promote engaged citizenship, and for local communities to overcome usually longstanding town/gown divisions. Academic linkages may have relevance here—respect for diversity may be an academic goal of a multicultural education course; nevertheless, of primary importance is the conceptualization of service-learning as fully analogous to the credo of the liberal arts to promote the engagement with alternative and enabling visions of self and society.

A political conceptualization of service-learning is focused on the promotion and empowerment of the voices and practices of disempowered and nondominant groups in society. Service-learning is seen to redress power imbalances, to legitimate marginalized communities and groups, and to harness institutional resources for social change. Service-learning as a pedagogical endeavor is here viewed as the embodiment and enactment of a social justice worldview, where the personal and the political meet in a substantive praxis. The seemingly neutral positioning of academics as distanced from and implicitly superior to the everyday world is critiqued as the colonial hegemony of male-centered theorizing. Instead, a political conceptualization of service-learning demands respect for the community, reciprocity of services, and a vision of higher education as a central agent of/in an equitable society.

A postmodern conceptualization of service-learning is focused on how the service-learning process creates, sustains, and/or disrupts the boundaries and norms by which we make sense of ourselves and the world. The emphasis is on the *how* of the experience, how service-learning works in its micro-practices: how, for example, we develop our notions of "servers" and "served," how power relations (between students and teachers or the college and community) are revealed or hidden, how we come to legitimate certain forms of knowledge and practice rather than others. By exposing the construction of such boundaries, categories, and norms, a postmodern conceptualization of service-learning works to disrupt the "commonsensical" and "natural" presumptions of our culture's grand narratives.

I suggest that traditional models of service-learning fundamentally privilege the cultural and political conceptualizations outlined above. Whether viewed from its historical origins (Stanton et al. 1999) or in its present-day arguments (Billig and Welch 2004), service-learning is committed to a vision of individual and social progress and change. For now (and I return to this in more detail in the next section) it is enough to point out that this is the classic liberal argument embedded within the modernist Enlightenment project. Whether taken up by philosophers such as Jurgen Habermas (1989) and Amy Gutmann (2004) or academic leaders in a "Declaration of Civic Responsibility of Higher Education" (Compact 2004), the key is the emphasized construction of a better world through an incremental progression achieved through rational and logical discourse and persuasion.

These are noble aspirations, and service-learning advocates seem to come to these perspectives with all of the "right reasons." What specifically interests me, though, are two related implications of holding such perspectives: first, that cultural and political conceptualizations are the default options for service-learning, and, second, that the technical "best practices" conceptualization is a very recent phenomenon. With regard to the former, it is important to note that service-learning arose within the crucible of the civil rights era and that this history informs almost all modern-day enactments of it. With regard to the latter, it becomes clear that positioning service-learning through a technical conceptualization (e.g., "service-learning enhances student outcomes") is a strategic move toward supporting institutionalization.

As the title of an influential book by Eyler and Giles (1999) suggests, "Where's the Learning in Service-learning?," service-learning advocates are only now beginning to talk the academic talk in order to gain institutional footholds and legitimacy (see also Zlotkowski 1995).

As the concluding section of this chapter makes clear, quantifying the "value added" has become an important (and problematic) component of demonstrating the worthiness of service-learning. For now it is enough to note that such data-driven agendas must be acknowledged as instrumental goal-oriented endeavors meant to support legitimization in the academy; what is also clear is that cultural and political conceptualizations of service-learning are foundational for the field. Which is, at least for Stanley Fish, exactly the problem.

A FISH(Y) SUGGESTION TO EMBRACE STANLEY?

My argument is that Stanley Fish rejects service-learning when it is conceptualized through cultural and political lenses. He argues that not only are liberal rationales of academic practice (e.g., "empowering," "fostering civic engagement") theoretically vacuous, they are internally self-contradictory and destructive of the actual task of teaching and learning within higher education. The key to understanding the power of his arguments vis-à-vis service-learning is to view cultural and political conceptualizations of service-learning as weak and strong versions of multiculturalism, respectively.

Weak versions of multiculturalism emphasize notions of cultural pluralism in order to promote a climate of respect, honor nontraditional habits of mind and practice, and treat others with the dignity and deference due within the diversity of our mosaic (read: *not* melting pot) culture. Strong versions of multiculturalism emphasize social justice in order to problematize claims to "objective" knowledge, dispute asymmetrical relations of power between dominant and nondominant groups, and advocate for more equitable opportunities and outcomes across economic, educational, and political spheres. These distinctions are sometimes labeled as difference multiculturalism versus critical multiculturalism (see Butin 2003b); in either case, these theoretical boundaries are long-standing and prevalent in fields such as education and anthropology.

Fish (1999) terms difference multiculturalism (which I from now on, for the sake of brevity if not clarity, conflate with cultural conceptions of service-learning) boutique multiculturalism:

> Boutique multiculturalism is characterized by its superficial or cosmetic relationship to the objects of its affection. Boutique multiculturalists admire or appreciate or enjoy or sympathize with or (at the very least) "recognize the legitimacy of" the traditions of cultures other than their own; but boutique multiculturalists will always stop short of approving other cultures at a point where some value at their center generates an act that offends against the canons of civilized decency as they have been either declared or assumed. (p. 56)

A multiculturalist who respects and accepts different opinions on abortion would be "stopped short" by attempting to respect and accept the blockage of abortion clinics and the harassment of the women who attempt to use it. Likewise between respecting and accepting different notions of women's rights and attempting to respect and accept some African groups' practicing of female genital mutilation. For a service-learning advocate, the breaking point would occur, for example, between working with families on child-rearing practices and supporting a family's desire to yell at and harangue their children as a form of discipline. Or between working with an immigrant-advocacy organization to teach new immigrants how to navigate the social welfare system and working with an immigrant-advocacy organization to preach the intolerance and racism of white folks.

The breaking point in all of these cases is, Fish argues, the inability of the multiculturalist/service-learning advocate to "take seriously the core values of the cultures he tolerates . . . [because] he does not see those values as truly 'core' but as overlays on a substratum of essential humanity" (p. 57). The multiculturalist is attempting to substantiate a liberal vision by reifying a notion of "universal humanness" that should overcome any and all dishar-monizing differences due to race, ethnicity, religion, language, socioeco-nomic status, and so on. The rationale is the Rawlsian notion of justice nicely crystallized in the "veil of ignorance" principle: that one should make governing rules as if one did not know in which social position (gender, race, socioeconomic status, etc.) one would be born into.

Thus, it seems logical to attack the blockade of abortion clinic, female genital mutilation, corporeal punishment, and hate speech. We wouldn't want all of a sudden, but for the grace of God, to be placed on the other side. Yet the implications of this line of reasoning is the utter dismissal of the most basic organizing principles of the other culture: religious fundamentalists (and others) *believe* in the sanctity of the unborn child; some African groups *believe* in the physical removal of the female libido; some families *believe* in strict adherence to respecting authority; some ethnic groups *believe* in the Satanic nature of the white man. To pick and choose which beliefs one is going to legitimate, with the added caveat being that one legitimates only those beliefs that one is already comfortable with, is actually the exact opposite of being multiculturalist; it is being a uniculturalist in multiculturalist clothing.

A strong multiculturalist perspective (much like a political conceptualiza-tion of service-learning) may appear as a means out of this dilemma. For a strong multiculturalist subscribes exactly to raising and respecting the voices of those different than oneself in order to support a higher principle of, for example, equal rights or free speech. Thus the ACLU supports both Orthodox Jews and the Ku Klux Klan's right to march. For Fish, this is because the strong multiculturalist "will want to accord a *deep* respect to all cultures at their core, for he believes that each has the right to form its own identity and nourish its own sense of what is rational and humane. For the strong multiculturalist, the first principle is not rationality or some other supracultural universal but tolerance" (p. 60).

But this tolerance only goes so far, for it too is predicated on a liberal notion of rational and logical persuasion of the other. Fish continues,

> The trouble with stipulating tolerance as your first principle, however, is that you cannot possibly be faithful to it because sooner or later the culture whose core values you are tolerating will reveal itself to be intolerant at the same core . . . Confronted with a demand that it surrender its viewpoint or enlarge it to include the practices of its natural enemies—other religions, other races, other genders, other classes—a beleaguered culture will fight back with everything from discriminatory legislation to violence.   (pp. 60–61)

This is strikingly clear in our post-9/11 world. Political conservatives effortlessly appropriate a hard-line rhetoric of limiting freedoms for all and expanding (often draconian) oversight of the (Muslim) Other. Political liberals, meanwhile, are left fumbling for an appropriate response that both demonstrates their commitment to American security without seemingly compromising their most basic tenants of respect and equal justice for all. The political Left is thus in a no-win situation, either looking soft on security or hypocritical to their values.

Open-mindedness (or "tolerance" or "equal rights") thus become rhetorical gestures that are empty of substantive meaning; either they simply reinscribe the position already assumed by the dominant culture (boutique multiculturalism) or they break down in limit cases where adherence to such principles pose a threat to the culture articulating them (strong multiculturalism). Richard Rorty (1991) has put it more caustically, suggesting that liberals have "become so open-minded that our brains have fallen out" (p. 203). Likewise, such seemingly universal principles are internally contradictory in that one's espousal of such principles is a pragmatic impossibility if one's aim is to be true to the principle.

Fish's position is rather to view all such articulations of principles as political strategic moves intended to win whatever battle one is fighting for. When one is fighting for affirmative action, for example, arguments of racist historical conditions and past suffering due solely to skin color are essential. When one is fighting against affirmative action, arguments of "reverse-racism" and "color blindness" are essential. And when one is fighting against the fight against affirmative action (which is what Fish is doing), arguments of the vacuity and internal contradiction of "reverse-racism" and "colorblindness" are essential. For Fish, there are no grounding universals; it is politics all the way down.

It is critical to understand that Fish's position is not just as a curmudgeonly contrarian determined to attack liberal and conservative positions to gain notoriety and air time and big speaking fees (though I'm sure that doesn't hurt). Fish's positions, I suggest, are directly related to his literary methodology. Fish first made his reputation by shaking up Milton scholarship and later literary theory more generally through close textual readings, his demand for a reader-response theory of literature, and his argument to focus on how a text works rather than on what it seemingly says (1967, 1972, 2001).

Fish's theoretical argument is that texts function as "self-consuming artifacts" (1972) in which "the reader is first encouraged to entertain assumptions he probably already holds and then is later forced to reexamine and discredit those same assumptions, [and this pattern] is repeated again and again" (p. 10). The text's goal, according to Fish, is to disrupt the reader's sense of certainty: "The strategy or action here is one of progressive decertainizing" (p. 384). Fish's goal is to understand how this happens: "For the questions 'what is this work about?' and 'what does it say?', I tend to substitute the question 'what is happening?' and to answer it by tracing out the shape of the reading experience" (pp. xi–xii).

The analysis of how a text works transfers the focus from a search for (often static and reductive) answers to an explication of the impact of the reading experience upon the reader: "An observation about the sentence as an utterance—its refusal to yield a declarative statement—has been transformed into an account of its experience (not being able to get a fact out of it). It is no longer an object, a thing-in-itself, but an event, something that *happens* to, and with the participation of, the reader. And it is this event, this happening—all of it and not anything that could be said about it or any information one might take away from it—that is, I would argue, the *meaning* of the sentence" (p. 386).

The implications are far-reaching. First, one no longer has access to "universal" notions of a "good" text; rather, all responses to a text are locally situated and thus locally influenced. Second, and following from the first, no method is value-free; Fish, for instance, offers that he is drawn to such "self-consuming texts," that "do not allow a reader the security of his normal patterns of thought and belief" (p. 409). Third, and following from the second, the decision to appropriate one literary method over another is a political move meant to overcome other political moves to support other literary methods.

Finally, and what will ultimately bring us back to the specific relevance for this chapter and for the theory and practice of service-learning more generally, Fish suggests that the point of all of this methodological theory is not to come to some final and ultimate point:

> Coming to the point [of a text] is the goal of a criticism that believes in content, in extractable meaning, in the utterance as a repository. Coming to the point fulfills a need that most literature deliberately frustrates (if we open ourselves to it), the need to simplify and close. Coming to the point should be resisted, and in its small way, this method will help you to resist.   (p. 410)

Fish's argument that academicians stick to questions of truth rather than civic engagement is but an expansion of this line of thinking. For Fish, truth is not some static and fixed object to be discovered but a process to be engaged in; to paraphrase Dewey's (1910) notion of mind, truth must be

viewed as a verb rather than a noun. Civic engagement (or openness to diversity or morality or whatever) is for Fish but a fixed principle pretending to
universalism. While well-meaning, such principles deny that they are the
product of particular readers with particular values attempting highly particular political moves. To claim "civic engagement" as a goal is to choose a
particular truth game without acknowledging the political move being made,
thereby both constraining discussion of the legitimacy of such a goal and
denying the opportunity to develop and articulate alternatives.

Fish's demand that we stick to teaching the truth is thus a profound
articulation for a (if you will) "self-consuming pedagogy," one that does not
allow the student the security of his normal patterns of thought and belief.
This is not about freeing higher education from politics; rather it is the
explicit acknowledgment that higher education is politics all the way down
(politics here understood, with Fish, as "subject to dispute"). And as such,
that it must stay open to the vigorous debate and discussion at the heart of
the "search for truth."

It is here, by the way, where the political Right falls off the theoretical
train. Fish's work is regularly cited by conservative causes such as Students
for Academic Freedom (www.studentsforacademicfreedom.org). They
believe they are citing him exactly because of his demands for higher education to stick to teaching the truth (here understood as a noun), to vigorous
debate of multiple (read: non-liberal) perspectives, and not to propounding
political and partisan bromides. Yet what Fish's work really shows is that conservative arguments simply repeat the exact rhetorical move he is tearing
apart: of raising false universal principles ("intellectual diversity," "affirmative
action for conservatives") for their political goals. Fish rightly argues that the
conservative argument is a Trojan horse because it argues for diversity as a
goal onto itself, rather than as a means (see Fish 2004b). Put otherwise,
there is no point in having "intellectual diversity" for the sake of intellectual
diversity. Intellectual diversity, for Fish, should simply be a means toward the
goal of a particular truth.

Fish's stance, it seems to me, is but a rearticulation of the postmodern
tropes of the incredulity to metanarratives, the deconstruction of the stable
modern self, the reading of texts for how they work rather than what they
purport to say, and the explication of the linkages between language, knowledge, and power. It is also, I suggest, an entrance into supporting (and
strategizing) for a more useful and defendable conceptualization of what
service-learning is and could be.

Service-Learning as
Self-Consuming Pedagogy

At the heart of this chapter's line of argumentation is that service-learning
can be viewed as a "text," and as such one of multiple classroom experiences
that are consistently "read" by students (see Varlotta 1997). This in fact
dovetails with Fish's arguments for a reader-response theory of literature, for

service-learning privileges the local situations and reactions of "readers" to the text. Service-learning advocates just term this reflection. To hew to Fish's argumentation, though, is to also accept that we must ask how the text works rather than focus on the "point" of the reading.

So what if we took Fish seriously and ignored the question of "What's the point of service-learning?," where "the point" is to be understood in its connotations of liberal notion of progress and fixed/hoped-for outcomes. Cultural and political conceptualizations focus almost exclusively on making points—about diversity, about social justice, about reciprocity—through the service-learning experience. What would occur if cultural and political conceptualizations of service-learning are bracketed? How would service-learning work? If service-learning is the text and the student is the reader, what does service-learning as self-consuming pedagogy do?

Let me suggest that, following Fish, it does four things. First, it subverts the notion that there are universal "best practices" of service-learning. Just as there is no objective way to claim the "goodness" of a text, there is no objective way to claim for the "best" way to do service-learning.

This is not a claim for radical relativism. Truth claims are always already situated within larger communities of practice, where experts make specific claims for and arguments about what is "accurate" and what is "good" with often highly refined and calibrated tools to judge such accuracy or goodness. Milton scholars make daily judgments on Fish's literary arguments; educational psychologists make daily judgments on students' learning gains; service-learning advocates make daily judgments on the value of a service-learning program for a community partner. As such, scholars make particular decisions as to how they will define good practices (e.g., number of reflection hours, number of "contact hours," what counts as learning) that may or may not support particular forms of service-learning practice. The point here is that these are intersubjective deliberations made from particular local perspectives with particular local goals.

The second implication, which arises from the first point, is that particular deliberations and articulations of the goodness or value of service-learning are always already culturally saturated, socially consequential, politically contested, and existentially defining decisions. There is no value-free articulation of what is good service-learning practice; as service-learning scholars and practitioners we make our judgments based on our particular goals and beliefs of who we are and why we do the things we do (see Birge, this volume).

The third implication, which stems from the second, is that these positions are developed, whether consciously or not, in contrast to competing definitions. Service-learning is a strategic move. It may be done (ideologically) to subvert the banking model of education as the imposition of inert content knowledge upon passive and empty students; it may be done (pragmatically) to enhance one's teaching of specific content matter; it may be done (politically) to enhance one's position as an advocate for the dispossessed. In all cases, service-learning is engaged in as a means to shift the normative assumptions of what counts as "good" teaching and learning in higher education.

Finally, what service-learning really does, to paraphrase Fish, like most good teaching, is deliberately frustrate (if we open ourselves to it) the need to simplify and close off our experiences. Service-learning, I argue, helps us to resist coming to premature closure. Academic subjects are inherently messy; our attempts to, for example, fashion linear and chronological syllabi around particular themes and topics is but the chimera of order overlaying the turbulence and turmoil of lived realities. Service-learning is but the working through of such issues within the context of a contained, academic, and reflective encounter.

Put otherwise, service-learning is a truly postmodern pedagogy in its incredulity to metanarratives (again, if we open ourselves to it). Whether higher education students are dialoging with prisoners, tutoring migrant students, or developing business ventures to support projects for a homeless shelter, they are constantly encountering the dilemmas and ambiguities of living with and through the complexity of how life works. To simply find an answer—"prisoners deserve their incarceration," "I am helping these uneducated youth," "We are making a difference"—is to foreclose a discussion through reductive psychological mechanisms.

Finding such answers is of course done all the time; we cannot and do not live always aware of and grappling with what William James so famously termed the "blooming, buzzing confusion." We create patterns, attribute meanings, develop habits. What becomes troubling, from a postmodern perspective, is when the performativity of being is displaced and hidden by the empty signifiers of "natural" or "commonsensical" or "that's just the way the world is."

Service-learning as postmodern pedagogy may thus be seen as a riposte against the foundationalism implicit in "commonsense" thought. It is a pedagogy immersed in the complexities and ambiguities of how we come to make sense of ourselves and the world around us. These are not easily embraced perspectives; service-learning engages in what I have termed "contested knowledge" (Butin 2005) in that content knowledge is viewed as potentially disruptive of a student's sense of self and thus to be resisted. As such, I want to more thoroughly discuss and embrace the implications caused by service-learning as a postmodern self-consuming pedagogy.

Let me frame this through a concrete example. I require the students in my social foundations of education course to tutor high-school Hispanic migrant youth through a local community organization. Sooner or later in the semester the discussion always arises as to the value of this endeavor; the youth, after all, seem not to understand much English, seem not to care much about homework or schooling, and seem not to be pushed by their peers or family to become successful. Some students will of course disagree, citing their particular experiences of the interested and interesting youth they are tutoring; for the (silent) majority of my students, though, the questioning hits home.

I then require my students to spend some of the tutoring time on the Internet on Spanish-language websites or sites featuring pictures of the youth's home countries (the tutoring occurs in the college's library and the students and youth have access to the computers). I also require them to attend one social and/or cultural event with the youth they are tutoring on campus and one in the youth's community (e.g., a sporting event at the high school/college; an art exhibit at the high school/college).

The class discussion after the completion of these activities is starkly different. Students talk about the youth's interest and energy and their desire to understand and be understood. We talk about how different contexts prompt different responses and how academic learning seems to lack much of what animates these youth. The lesson is seemingly learned: in different situations youth who are apparently disengaged and disinclined become passionate, articulate, and intelligent. This is a lesson enough within teacher education committed to a multicultural vision of tolerance and plurality. But I push on.

I ask my students how they felt being in a different context (the high school) and outside of their normal role as tutors? I prod them to explain what the websites were about. I ask them who was teaching whom? The surety of their responses diminishes. At first a simple binary takes hold: "the youth were the teachers and we were the students"; "they understood the websites and we didn't." But such (inverted) binaries do not hold. What instead becomes revealed are the implicit assumptions and norms they carried into the service-learning experience about doing good, about "serving," about the value and method of teaching, and about the notion of youth as passive and thankful subjects.

The service-learning experience has become self-reflexive. Or as Fish would phrase it, self-consuming. Every single further encounter with the migrant youth must be read through the lens of our discussion about our roles and our assumptions of what we do and why. When we help these youth with their homework, is it for their benefit or ours? Why do we feel good about doing this? Will we use this as a discussion point with our friends later that evening? Will we put it on our resumes? Will we realize how much of an impact we can have? Will we ask why they don't learn English? Why school is boring? Or will we stay silent? And if we do, why?

The point here is not that there is a final point to the questions, some "right" answer that either my students or the migrant youth will come up with. My point is that the service-learning experience has forced my students to realize that what they do and think is not "natural." They may still leave my class believing whatever it is they believe; that they were ultimately helpful to the migrant youth, teaching them English and respect for education; or that they are now much more cynical of an educational system unable to accommodate the needs of diverse learners. But whatever performative move they make must now be read through the experience of the performance.

This is not to raise the specter (yet again) of a radical relativism with no basis for discerning what counts as good work, good tutoring, good service-learning, good whatever. My students must be able to explain themselves to me and to their peers through their discussions, their papers, their journals, and their actions. It is this intellectual community that offers, as always, the intersubjective boundaries of what counts as "good" (Fish 1980). In the end, of course, I cannot control what they leave class believing; all I can do is offer as many experiential intersections by which they come to explicate and reflect on the nature of their performative utterances and beliefs. All I can do is promote a "weak overcoming" that always already questions the foundations we hold as we go about constructing our foundations.

IMPLICATIONS FOR THE SERVICE-LEARNING FIELD

If service-learning is truly a postmodern pedagogy, we may still ask, "So what?" Let me suggest that there are three direct implications for the service-learning field. The first is that viewing service-learning from a postmodern perspective clarifies the resistance (implicit and explicit) from some students, faculty, administrators, and policymakers. What becomes clear is that service-learning is a partisan practice, especially when it is enacted through cultural and political conceptualizations. I do not mean this in a derogatory sense; all pedagogical practices are partisan. I mean it rather as clarifying the language games of a practice that claims to universality. It thus becomes clear why some students seemingly "don't get it" (Jones et al., this volume); they resist the service-learning experience because it contradicts their sense of the world and their political orientations to it. This is a politically problematic situation for the service-learning field because it cannot claim to simply a neutral teaching practice.

This can most clearly be seen by the level of serious attention received by David Horowitz's "Academic Bill of Rights" (http://www. studentsforacademicfreedom.org/abor.html). Working through his Center for the Study of Popular Culture, Horowitz has spent years arguing that the liberal bias of higher education is depriving students of a fair and balanced education: "You can't get a good education if they're only telling you half the story," argues Horowitz. His research shows that liberals outnumber conservatives almost ten to one in higher education (Horowitz and Lehrer, n.d.); from this have sprung numerous legislative initiatives (with Colorado set to vote this year) calling for, in effect, affirmative action for conservative-leaning academics.

Service-learning has not yet been lambasted; but it cannot forever stay out of the conservative crosshairs. Traditional service-learning practices promote highly partisan orientations toward, for example, what it means to help the needy, what this help might look like, and who should be at the forefront of such societal changes. To put it bluntly, service-learning practices do not usually promote a perspective of trickle-down economics favored by neo-conservatives. While viewing service-learning from a postmodern perspective does not immediately make the problem go away, it nevertheless highlights the fact that this is not a universal and neutral way to "do" higher education. In so doing, it puts the issue squarely on the table to be debated.

A second implication of service-learning as postmodern pedagogy is that it allows faculty to embrace the pedagogical potential of service-learning. Fish (2003) provocatively asserted that, "You have little chance however (and that entirely a matter of serendipity) of determining what they [students] will make of what you have offered them once the room is unlocked for the last time and they escape first into the space of someone else's obsession and then into the space of the wide, wide world" (C5). This is not a stance of dispirited resignation. This is, for Fish, the reality that texts stop working once they are not being read. Which is not to say that texts are irrelevant.

The most powerful texts, the "self-consuming artifacts," in fact have such a profound affect that the reader can no longer return to the starting point where she began—before she started reading. Self-consuming texts subvert a return to the origin; Fish's consistent use of Wittgenstein's metaphor of a ladder with steps falling away as one climbs up it ("[the reader] must, so to speak, throw away the ladder after he has climbed up it," Fish quotes Wittgenstein) is deeply deliberate. The self-consuming text leads the reader in such a way as to subvert any means by which the reader can return by the way that she came. She must instead live with the consequences of having read. Even rejecting the reading does not negate the experience; it simply marks the next move in the experience of reading.

Service-learning is analogous. Service-learning is a fundamentally embodied process. As such, students cannot simply engage in an intellectual exercise. They must embrace, whether consciously or not, the actions within the experience because they are actors within it. Service-learning experiences can thus be viewed not as attempting to make a point, but to actually be the point.

This leads to the final, and that I find to be the most pressing, implication of service-learning as postmodern pedagogy. Namely, that service-learning seems verged on accepting a highly problematic Faustian bargain of gaining institutional legitimacy by giving up its transformational opportunities. At the crux of this bargain lies the quantification of service-learning.

The service-learning field is deeply engaged in questions of institutionalization (see Hartley, Harkavay, and Benson, this volume). A major component in this endeavor has been the emphasis on the quantification of the impact of service-learning upon specific outcome measures. Service-learning advocates consistently point to quantitative studies supporting the role of service-learning in enhancing, among other things, content knowledge acquisition, civic engagement, and openness to diversity. This data-driven advocacy seems almost intuitive; service-learning proponents are attempting to cross a legitimacy threshold whereby faculty and administrators come to view service-learning as simply one of multiple pedagogical strategies available for faculty. Hard data appear a straightforward path to achieve such legitimacy. Thus for example, if the data reveal that service-learning work with terminally ill AIDS patients enhances students' thinking about the health care system or about issues of death and dying, then faculty can simply move forward with service-learning as just another pedagogical tool on their methodological tool belt.

I suggest, though, that service-learning advocates have been playing the wrong political game. To see why, one need only reverse the scenario. If, instead, I found that working with terminally ill AIDS patients actually impaired students' understanding of social health policies, would service-learning advocates desist their work for fear of worsening the academic decline of our college students? I doubt it. Rather, I imagine they would question the validity and reliability of such data: How is "understanding" being measured? Is success defined instrumentally (i.e., test grades) or holistically (i.e., emotional intelligence, long-term changes)? What was the timeframe of my assessment procedures? Did I use pre- and post-tests or interviews? Was there an adequate control group?

The point here is that advocates and skeptics of service-learning have each constructed, a priori, their respective criteria of what constitutes success before the first scrap of data has even been collected. As such, the collection of seemingly objective, quasi-experimental data by service-learning advocates is done for the sole goal of convincing those other academics; contraindicative data collected by service-learning advocates result only in the bemoaning of the coarseness of poorly defined proxy variables. Likewise, no amount of hard data is going to convince so-called academic purists that going out in the community to do good is somehow going to positively impact students' learning of, for example, Milton's poetry.

What I am suggesting, therefore, is that service-learning may need to play a different political game to promote and sustain service-learning in higher education. Women's studies and qualitative research, for example, did not become a part of the academy by playing by the prescribed rules. At stake is whether service-learning plays by the normative "old school" rules of the academy as a neutral and objective discussion about truth or attempts to modify what counts as the rules (and truth) in the academy.

Put otherwise, the way to approach this issue is to view service-learning as any other attempted pedagogical strategy vying for normative sovereignty within the academic landscape. To insist on quantification as academic corroboration privileges the legitimacy of the academy as an objective, rationally bounded enterprise. Problematically, though, such reliance on the quantification of service-learning ends up creating strong, normative, and police-able boundaries of what can and cannot constitute service-learning. The service-learning field, in other words, becomes bound by an academic interpretive community not of its own making. This would be like, analogously, qualitative studies beholden to quantitative norms; or women's studies operating by male-centered models of practice.

In the end, it seems to me, Stanley Fish offers us a way out (and forward). By acknowledging service-learning as a partisan practice, by embracing the transformational potential of a postmodern pedagogy, the service-learning field can continue to experiment with different notions of what works in service-learning and how it works. To quantify the "best practices" of service-learning is to domesticate it, much like a "definitive" reading of Milton attempts to do. Fish resists this to the end, for the reader will always make her own meanings from the experience; and, if done carefully enough and with enough courage, she will find that the experience has forced her to rethink why she was doing the experience in the first place. The text deconstructs from the inside out.

Service-learning as a self-consuming pedagogy, as postmodern practice, is for me a liberating stance. It doesn't ask me to make my students into civic minded citizens, it doesn't ask me to solve the social justice issues facing me locally and globally, and it doesn't force me to combat (or embrace) globalization. These may of course occur as byproducts of the service-learning experience. But I cannot expect it; much less should I attempt to force it. What instead I can and should do is look carefully at the practices in my own shop—the enactment of service-learning experiences—and figure out how it works. This is, it seems to me, more than enough to ask for.

REFERENCES

Billig, S. & Welch, M. (2004). *New perspectives in service-learning: Research to advance the field*. Greenwich, CT: Information Age Publishing.

Bringle, Robert & Julie Hatcher (1995). A Service-learning curriculum for faculty. *The Michigan Journal of Community Service-Learning*, 2: 112–122.

Butin, D. W. (2003a). Of what use is it? Multiple conceptualizations of service-learning in education. *Teachers College Record*, 105(9): 1674–1692.

Butin, D. W. (2003b). The limits of categorization: Re-reading multicultural education. *Educational Studies*, 34(1): 62–70.

Butin, D. W. (2005). Identity (re)construction and student resistance. In D. W. Butin (Ed.), *Teaching social foundations of education: Contexts, theories, and issues* (pp. 109–126). Mahwah: NJ: Lawrence Erlbaum Associates.

Campus Compact (2004). Presidents' declaration on the civic responsibility of higher education. Available at http://www.compact.org/presidential/declaration.html. Accessed 10/27/04.

Dewey, J. (1910). *How we think*. Boston: D.C. Heath & Co.

Eyler, Janet & Dwight E. Giles, Jr. (1999). *Where's the learning in service-learning?* San Francisco, CA: Jossey-Bass.

Fish, S. E. (1967). *Surprised by sin: The reader in "paradise lost."* New York: Macmillan; St. Martin's Press.

Fish, S. E. (1972). *Self-consuming artifacts: The experience of seventeenth-century literature*. Berkeley: University of California Press.

Fish, S. E. (1980). *Is there a text in this class?: The authority of interpretive communities*. Cambridge, MA: Harvard University Press.

Fish, S. E. (1994). *There's no such thing as free speech, and it's a good thing, too*. New York: Oxford University Press.

Fish, S. E. (1999). *The trouble with principle*. Cambridge, MA: Harvard University Press.

Fish, S. E. (2001). *How Milton works*. Cambridge, MA: Belknap Press of Harvard University Press.

Fish, S. E. (2003). Aim low. *Chronicle of Higher Education*, 49(36): C5.

Fish, S. E. (2004a). Why we built the ivory tower. *New York Times*, 5/21/04, A23.

Fish, S. E. (2004b). "Intellectual diversity": The Trojan horse of a dark design. *Chronicle of Higher Education*, 50(23): B13–B14.

Fish, S. E. & Veeser, H. A. (1999). *The Stanley Fish reader*. Malden, MA: Blackwell Publishers.

Furco, Andrew (1996). *Expanding Boundaries: Serving and Learning*. Washington, DC: Corporation for National Service.

Furco, Andrew & Shelley Billig (Eds.) (2002). *Service-learning: The essence of the pedagogy*. Greenwich, CT: IAP Press.

Gutmann, A. (2004). *Why deliberative democracy?* Princeton, NJ: Princeton University Press.

Habermas, J. (1989). *Jürgen Habermas on society and politics: A reader*. Boston: Beacon Press.

Horowitz, D. & Lehrer, E. (n.d.). Political bias in the administrations and faculties of 32 elite colleges and universities. Available at http://www.studentsforacademicfreedom.org/reports/lackdiversity.html. Accessed October 28, 2004.

Kendall, Jane (Ed.) (1990). *Combining service and learning: A resource book for community and public service*. Raleigh, NC: National Society for Internships and Experiential Education.

Lisman, C. David (1998). *Toward a civil society: Civic literacy and service-learning*. Westport, CT: Bergin & Garvey.

Liu, Goodwin (1995). Knowledge, foundations, and discourse: Philosophical support for service-learning. *The Michigan Journal of Community Service-Learning*, 2: 5–18.

Michigan Journal of Community Service-learning (MJCSL) (2001). *Service-learning course design workbook*. Jeffrey Howard (Ed.). Ann Arbor, MI: OCSL Press.

Morton, Keith. (1995). The irony of service: Charity, project, and social change in service-learning. *The Michigan Journal of Community Service-Learning*, 2: 19–32.

Rorty, R. (1991). *Objectivity, relativism, and truth*. Cambridge: Cambridge University Press.

Sigmon, Robert (1979). Service-learning. Three principles. *ACTION*, 8(1): 9–11.

Stanton, T., Giles, D., & Cruz, N. I. (1999). *Service-learning: A movement's pioneers reflect on its origins, practice, and future* (1st ed.). San Francisco, CA: Jossey-Bass Publishers.

Varlotta, Lori E. (1997). A critique of service-learning's definitions, continuums, and paradigms: A move towards a discourse-praxis community. *Educational Foundations*, Summer: 53–85.

Wittgenstein, L. (2000). *Tractatus logico-philosophicus*. London, New York: Routledge.

Zlotkowski, Edward (1995). Does service-learning have a future? *The Michigan Journal of Community Service-Learning*, 2: 123–133.

SECTION II

TRANSFORMATIVE MODELS OF SERVICE-LEARNING PRACTICE

7

The Evolution of a Community of Practice: Stakeholders and Service in Management 101

Jordi Comas, Tammy Bunn Hiller, and John Miller

> What makes information knowledge—what makes it empowering—is the way in which it can be integrated within an identity of participation. When information does not build up to an identity of participation, it remains alien, literal, fragmented, nonnegotiable.
>
> Wenger 1998, p. 220

Setting the Stage

Sandy enters the Forum—a large lecture hall—on the first day of her introduction to organization and management class, MG 101. She stares up at more than a hundred fellow students. Having heard rumors of the hard work this course entails, Sandy nervously seeks out friends from her sorority to sit beside. As she heads toward her friends, an upper level student teaching assistant directs her to sit in her "company," one of four sections of about 30 students who will complete three projects together: a service project, a business project to fund the service, and a reporting project to document and interpret the experiences of both. Sandy moves to sit in Company C. Within four weeks, she and her fellow company members decide to create a new after-school computer lab at a local community center, funded by the sale of 400 custom-designed soccer scarves. The process of selecting their service and business projects has forced Sandy and her company members to brainstorm project ideas, research alternatives, advocate for a position, form coalitions, and reach a final decision. The next several weeks are spent designing their company organization, assigning jobs, planning their operations, carrying out their projects, and managing the dozens of unexpected issues and problems that arise. Sandy and her colleagues have both the freedom to experiment and explore how their company will operate, as well as frequent feedback and counseling from community stakeholders, teaching assistants, and professors. Within 14 weeks, Sandy and her colleagues have sold all their scarves, raised $3,000, purchased and solicited donations of computer equipment, organized, assembled, and decorated the lab, mapped out a training program for community center staff, begun an after school tutoring program utilizing the new computer lab, and reported the story of their experience, along with their interpretation of that story, to multiple stakeholders in multiple ways. This is a fairly typical experience of the MG 101 service-learning course, and one shared by about 6,100 management and non-management major undergraduate students forming 191 companies in the last 25 years. By the end of spring term, 2004, 191 MG 101 companies had given more than 51,500 hours of labor to service clients and provided materials and donations valued in excess of $211,000—making significant differences in the lives of children, adults, and improving the physical environment of the regional community.

MG 101 is an experiential general education course, committed to a balanced focus on liberal education and on preparation for professional careers in collaborative enterprises of any kind. Two key assumptions have guided MG 101's development. One, MG 101 grows from a belief that praxis (integrating theory and practice) is inherent in all human action that unfolds through myriad forms of collectivity. Two, MG 101 also grows from the need to instill stakeholder managing in our students and fellow citizens as a necessary remedy to the misguided visions and destructive consequences of seeing managing as solely instrumental or wealth-maximizing (Miller 1991). MG 101's ambition is to take experiential education principles to their logical conclusions, to provide a single, coherent, relevant, and integrated introduction to management theory and practice . . . in a one semester course . . . for everybody. In the words of one MG 101 student, MG 101 provides "instant street smarts."

When the original faculty designed and developed the MG 101 class in the early 1980s, neither service-learning nor social learning theory were particularly salient ideas, though the building blocks were certainly there (especially in discussions of experiential learning). By the late 1970s, there were increasingly widespread, critical debates (and a few interesting models) of alternatives emerging to both traditional pedagogy and models of management education derived from enormously successful, lucrative, and visible MBA programs anchored in narrowly instrumental, functional, and economist approaches (Kolb 1984; Livingston 1971). Those emergent alternatives came together in the MG 101 class and continue in its on-going exploration of alternatives to traditional learning and conventional managing (Miller 1996).

We think that MG 101 has much to tell us about critical approaches to management and about service-learning. The need is abundantly clear. Multiple and profound examples of the collapse of professional integrity throughout the business world over the last 20 years—the insider trading scandals, the failure of the S&L industry, and most recently the Enron bankruptcy and the various other scandals of "earnings management" and insider stock sales that littered the post dot-com bubble bursting—amply demonstrate that the time is right to revisit management education. The persistent lack of confidence among the public in businesses and managers are the troubling legacy of these colossal betrayals of the idea that management is a professional practice, not merely latter day robber-baron self-enrichment (Korten 2001; Tyson 2004).

How can management education confront the betrayal of professional practice? Because doing management in a dehumanizing, technocratic way has become so prevalent in natural communities of practice, a conscious and forceful effort has to be undertaken to do management in a way more responsive to the main criticisms of managing and management education (Samuelson 2000). In their introduction to a volume on service-learning and management education, Godfrey and Grasso write: ". . . a key component of the 'managerial mind-set' prevalent in American Business today originates in, and is perpetuated through, *business school education*. Business schools do more than merely train technical professionals in management; they also communicate a set of values regarding economic rationality and human worth that become the foundation of the managerial mind-set" (2000, p. 1, emphasis added). One solution, favored by many management academics, is to anchor management education firmly in the stakeholder theory of management and the firm. The crux of stakeholder theory is that "each stakeholder group has a right not to be treated as a means to some end, and therefore must participate in determining the future of the firm in which they have a stake" (Freeman 1984 in Donaldson, Werhane, and Cording 2002, p. 39). Because service-learning and stakeholder managing are so linked, "service-learning personalizes the concept of the 'stakeholder' and helps students see the legitimacy of each competing stakeholder's claim" (Godfrey 2000, p. 30). Bucknell's MG 101 course is one long-standing example of these efforts to make stakeholder managing a substantive, engaging, and moral endeavor for anyone.

In 25 years of evolving, MG 101 has not swayed from the deceptively simple premise that one must do management to be able to learn management. The converse applies as well: how one does management affects what is learned. MG 101 companies are unique because they are purposely double-bottom-line companies. Interest in this form of company has been exploding recently (Bornstein 2003). The double-bottom-line structure means that companies are challenged to maximize outcomes for two bottom lines—social impact and profit. This is also known as social return on investment, or SROI (see www.gettysburg.edu/~dbutin for a list of resources on social ventures and double-bottom-line companies). This structure has a host of implications for the learning experiences of students: the consequences of their company are real, they interact with a diverse array of stakeholders, and they develop an identity and practice of stakeholder managing. After 25 years, MG 101 has acquired rich institutional history and well-tested collective routines. Despite this institutionalization and the confidence of longevity, each session of MG 101 is imbued with a sense of novelty and discovery. This sense of discovery comes from the other key element of the design for learning (aside from the double-bottom-line structure): the design relies on students to form and learn from their own communities of practice. To see how this works, we will look at MG 101 through the lens of social learning theory and its concept of communities of practice (Wenger 1998).

Adopting a communities of practice perspective remedies a tendency to think of service-learning in purely procedural terms. Service-learning is often thought of in somewhat mechanical or procedural terms: a class in which community service earns credit. This is akin to describing a symphony by its title. What matters more is the experience of the symphony, or, in this case, the class. Designing or reanalyzing a service-learning class or activity with an eye toward how practice develops and how people belong to the community of practice is a useful exercise. An educator wishing to do service-learning should realize the inherent dualities in any design for a class, in the scaffolding for learning. We will share what we have learned as educators from MG 101 about how to design for communities of practice.

We bring two sets of perspectives to the task of unpacking MG 101. One is that of the professors who have designed the curriculum and interacted with students as they traverse that curriculum and classroom. The other is that of a partial outsider who has observed MG 101 classes and labs, read student papers, and conducted research into how social networks and organizational knowledge emerge in the MG 101 companies.

SOCIAL LEARNING THEORY

Social learning theory describes how people learn as part of developing a practice, as opposed to learning as a special category, divorced from engagement in a world of real consequences. Social learning theory rests on three assumptions: (1) we are social beings; (2) knowledge is a matter of competence with respect to valued enterprises; and (3) knowing is a matter of participating in the pursuit of such enterprises (Wenger 1998, p. 5). As Wenger describes the claims-processors he studied: "Their learning is not merely a context for learning something else. Engagement in practice—in its unfolding, multidimensional complexity, is both the stage and the object, the road and the destination." Learning is developing competence in a valued enterprise, with the additional insight that both competence and what is valued are socially negotiated categories—they are entangled with their own past and with the structure of the social locations in which they are embedded. The social aspect of learning is described by Brown and Duguid (2001) as how

> people do not simply learn *about*; they also learn, as the psychologist Jerome Bruner suggests, *to be*. Learning, that is, doesn't just involve the acquisition of facts about the world, it also involves acquiring the ability to act in the world in socially recognized ways. This last qualification, "in socially recognized

ways," acknowledges that it is not enough to claim to be a physicist or a carpenter; people, particularly other physicists or carpenters, have to recognize you as such. (200)

As the quotation above explains, social practice and identity are linked. Practice is shared among individuals; practice shapes the identities individuals carry from one moment of practice to another. As Wenger describes, learning is "not just an accumulation of skills and information, but a process of becoming—to become a certain person or, conversely, to avoid becoming a certain person. . . . We accumulate skills and information, not in the abstract as ends in themselves, but in the service of an identity" (215). As management educators, we care about practice and identity because we seek for our students to develop alternative practices of managing and identities of the manager. For example, most of our students come into the course believing that leadership correlates perfectly with hierarchical position. We offer this reflection on what one student learned after his initial disappointment in not being voted the CEO of his company.

> Community is just as important as, if not more than, effectiveness and efficiency. Before, I had always thought, "Who cares what everyone thinks, as long as the goal is accomplished and in the best way possible!". . .
>   I learned that initiative and personal power do not guarantee formal authority. The two can even be inversely related, as in my case. An organization is not made through one person's effort; it must be grounded in the cooperation of many motivated people striving for the same goal. (Ben, Spring 2002)[1]

From this discussion of practice and identity, we see that learning can be reimagined as increasing membership in a community defined by a particular practice. Neither practice nor membership need necessarily conjure up normative expectations of greater empowerment or workplace democracy. A practice may be deeply marked by hierarchical power dynamics and marginalizing engagement by members. The achievement of social learning theory is to describe the social dynamics and local history of learning by doing. These dynamics are crucial for student experiences in a service-learning class. But, as educators, how can one plan for learning by doing?

DESIGNING FOR COPs

The introduction to this volume describes how service-learning brings students into the "messy, complex" world beyond the walls of the classroom. Clearly, the double-bottom-line structure of the students' companies in MG 101 brings them into that world. By designing the learning experiences to be about learning from their own experiences in their own communities of practice, MG 101 makes the world inside the classroom messy and complex and it invites students to explore and learn from that rich mess. Our pedagogical objectives extend familiar service-learning goals—to "make a difference in the community," to "learn to stand in the shoes of others," to "develop civic literacy" (Lisman 1998)—to focus on organizational and managerial skills, attitudes and ideas. For our students, "doing service" is undoubtedly motivating and enlightening; for us, however, the more fundamental and critical source of learning is the *process of managing* those service goals—organizing and developing a community of practice. We suggest that much can be gained from thinking pedagogically about the organizing that students do to improve both bottom lines as a crucial source for learning and not just the means to achieve their stated ends. One example comes from this student's comments on leadership.

> Many times I had tunnel vision, believing that my ability to be an effective leader was achieved by the end product, a sound set of financial statements to turn in. However, it is not the end product that is important when leading a group of people, but rather the journey along the way. By finding the balance of control versus delegation, acknowledging that part of managing a team is both recognizing when individuals are not performing well, but more importantly why, and how they can improve their performance, and setting an example which others will respect, I have come a long way in understanding what it takes to be an effective leader.   (Brandi, Fall 2000)

In Brandi's reflection, we see her identity as a leader shift from managing for a single bottom line to the idea that effective practice involves deeper interactions with other stakeholders in a collective effort.

We can easily imagine an attempt to duplicate MG 101 in another setting in which a lack of attention to designing for communities of practice could reinforce the practice of management as unnecessarily hierarchical, obsessed with egoistic power, unhealthy competition, and a dehumanizing set of control procedures. Because this type of management practice is so prevalent in business and society, a new community of practice could blindly recreate the practice despite the use of an ostensibly radically different approach. The key is how the class accounts for the formation of communities of practice. MG 101, through the professors' statements and its course manual (Miller and Hiller 2004), offers such an approach. The degree of success in affecting the practice of double-bottom-line goals and stakeholder managing has as much to do with what communities of practice emerge as with the written goals. Even being a service-learning course would not necessarily guard against a straightforward reproduction of management hegemony. The companies could see themselves as the contributors of superior knowledge and financial resources to deserving charities; the company members would draw the most meaning from beating the other companies on money raised; and the service projects would avoid any honest dialogue and power sharing among company members and community partners. The full potential of MG 101, or any service-learning course, is not to be found in reading the curriculum or the other artifacts left behind. If we are to achieve the kind of profound learning that is the aim of both critical learning and service-learning, then we must understand and constantly reexamine how the experiences of a service-learning class translate into emergent communities of practice and the social learning of a student contributing to and caught up in that practice.

How can educators, while acknowledging the emergent, grounded, and local qualities of communities of practice still do what we do, namely, teach? Wenger provides some starting points. The key is to understand how a class, a design for learning, is also a design for practice. And, as such, that design can not "attempt to substitute for the world and the entire learning event. It cannot be a closed system that shelters a well-engineered but self-contained learning process" (275). Despite how we talk about and imagine service-learning, Wenger's advice still challenges deep notions about the role of the educator. He writes:

> Learning cannot be designed. Ultimately, it belongs to the realm of experience and practice. It follows the negotiation of meaning, it moves on its own terms. It slips through the cracks; it creates its own cracks. Learning happens, design or no design.   (Wenger 1998, p. 225)

Putting together these two observations—that one designs for an open system with the awareness that learning derives from experience and practice, Wenger suggests that design be thought of as an architecture for learning (whether in organizations or schools). Ironically, this is the same idea we have used in explaining the approach in MG 101: the course builds a scaffold for learning. If we accept that learning by doing needs an architecture or scaffolding, then we must acknowledge that learning through communities of practice cannot be tightly controlled.

THE DESIGN FOR LEARNING IN MG 101

Each section of the course, containing about 30 students, forms a real organization—a MG 101 company—with real service commitments and clients, real business products and customers, real jobs, real money, real conflicts, real complications, real challenges, real results, real pain, and real joy. The consequences of their decisions and actions are real—not imaginary or speculative or hypothetical. Through the remainder of this chapter, we discuss several key elements of MG 101's design for learning. These elements include the promulgation of a group formation sequence, the integration of ongoing reporting linking service-learning to stakeholder management, the three projects, and the three themes of the practical theory curriculum.

SIX STAGES

As figure 7.1 demonstrates, MG 101 is structured so that each company goes through six developmental phases, building on Tuckman's (1965) model of group development. Encouraging the students to be consciously aware of these phases as they live through them helps them to be more intentional about their own communities of practice. Learning through increasing membership

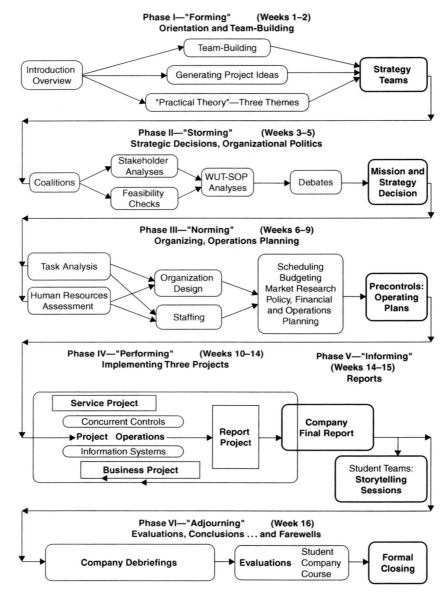

Figure 7.1   MG 101 Project activity timeline
Source: Miller and Hiller, 2004, p. 15.

in a community of practice takes time and unfolds in different modes of belonging—through engaging the community, through imagining one's identity in relation to the community, and through aligning practices and identities (Wenger 1998). In MG 101, planning for the phases and the changes in rhythm in each phase creates various facilities for belonging to the community.

In the forming phase, students begin to establish the bases of friendships (or the contrary) with their new colleagues. Quickly, relative strangers get acquainted, learn what they might expect from each other individually and collectively, and begin to think of "us" as a single, coherent entity, different from "them." During the storming phase, each MG 101 company decides on service and business projects, and develops a mission and strategy statement and a company name reflecting those decisions. These decisions constitute commitments to many different stakeholder groups, both directly, in terms of the specific activities and products they will deliver, and implicitly, in terms of the quality standards with which they will deliver them. The work in this phase reflects an increasingly intense political process, during which the company's power structure begins to emerge. Individuals engage through strategy teams and coalitions, they imagine projects and missions, and they align their various ideas and tentative plans to ideas of competence that come from the professors (e.g., competence as inquiry-based decision-making) and to more emergent ideas of success (like making a real difference, making a large profit, etc.). In this emotionally dynamic stage, the companies are on an accelerated course for developing their practices of managing and their identities by virtue of experiencing the several modes of belonging. The process of working through disagreements can, if managed well, produce the clarity about group goals and high levels of commitment necessary for moving to the norming phase.

Given commitment to common objectives, members then work to figure out who will do what, how individual efforts will be coordinated, and which of the many potential actions of members will be excluded from collective activity. Because it is concerned with defining and establishing machinery for enforcing shared expectations of individuals' behaviors, this phase is called "norming"—even though that name overemphasizes conformity and discounts critical tasks of dividing labor, assigning tasks to specialists, and delegating formal authority for individual discretion. The norming phase concludes with presentation of a company's operating plans to the MG 101 Board of Directors for loan approval (and resulting celebrations)—a fine example of an aligning moment.

In the next phase—performing—company members actually carry out the missions to which they committed themselves in the storming phase and planned how to accomplish in the norming phase. The performing phase is crucial for the practice of managing to evolve. Because the goal decisions of the storming phase and the job assignments and other plans of the norming phase will inevitably have been based on imperfect predictions about the real state of affairs in the performing phase, really successful performance also involves "winging it"—creative, spontaneous adaptation, initiative, and leadership. At a basic level, MG 101 is learning by doing, but at a deeper level, it is doing based on the emergent community of practice, including the double adaptation of adapting to conditions while adapting to the community of practice. Doing in MG 101 is negotiating meaning on a terrain of common and contested meaning.

Because MG 101 company members are collectively and individually accountable to multiple stakeholders for the results of their decisions and actions (unlike the subjects of Tuckman's group development observations), and have the responsibility to tell themselves, each other, and various external stakeholders about what they have done, MG 101 has defined (as its own community of practice!) a distinct fifth organization development phase—informing. The informing phase entails collaborative efforts to sort through and assemble, then prioritize and edit, then integrate and tell what's worth telling—in ways carefully designed to make good sense to all appropriate audiences.

Finally, in the adjourning phase, members celebrate individual and collective accomplishments. They remind themselves of what they do and don't wish to bring with them to their futures—including relationships with other members. And they acknowledge their ambivalence about disbanding—the pain of farewell and relief of being done. The final two stages afford much imagining and aligning of the recently produced, local history of the company. The imagining and aligning that transpires, particularly during celebratory transitions between phases, reify the company as a real and consequential experience and not merely a "class."

REFLECTION AND REPORTING

Reflection on experience is a final stage of the companies' development. But individual and group reflection and ongoing learning are designed into the course to unfold from the beginning. Storytelling and writing are the most powerful integrating mechanisms in MG 101—between theory and practice, between plans and accomplishments, and above all, among the student company members. Students engage in an integrated sequence of individual and collective storytelling and writing assignments that anchor learning in coherent descriptions and interpretations of their experiences in the course. Periodic storytelling and writing assignments challenge the students to focus on their own experiences—to think about the relationships and patterns among events they are experiencing, to test the relevance of course concepts and methods to solving their problems, and ultimately, to make sense of and learn from their MG 101 experiences.

At key times in the company's development, individuals write learning reports in which they describe critical turning points in their experiences, analyze and interpret their experiences through multiple conceptual lenses, and think through how they can put that learning into action in the next phase of the course. The work of the norming phase—creating an organization design, job descriptions, schedules, budgets, and detailed plans for accomplishing their missions—provides key opportunities to do group-level reflection. These collaboratively produced operating plans are presented to the Board of Directors of MG 101, adding a valuable public quality to their reporting. After completing its mission, each company presents, to an even larger audience of stakeholders and classmates, a theatrical oral presentation that involves everyone in telling and interpreting the company's unique story. This story is also told in writing, through a comprehensive detailed company archives and a concisely focused company summary. Through a combination of their oral and written reporting, students are challenged to tell a complete, coherent, honest story of their unique experience, and demonstrate how the company learned about and integrated core course concepts.

The aim of this continuous reflection and reporting is for students to learn to pay attention to their own actions, and (often, for the first time in an academic setting) to take themselves seriously as sources of valid data for learning. They are encouraged to both be in the moment and to watch the moment. As described by one company's Chief Executive Officer, the students leave the course having learned how to learn from their own experiences—arguably, the most critical management skill.

> The models for organizations in readings and lectures are exactly that—they are models, or "grand narratives," not necessarily descriptions of anybody's personal experience. Storytelling is a way to learn from theories, especially when you realize that they won't exactly portray what happens to you. It's the easiest and most meaningful way to see how day-to-day events hang together, but it's especially useful in getting you to pay attention to things that happen that *don't* match the game plan. The surprises are always what make stories really worth reading. As I look back on my MGMT 101 story and see some of the places where there was conflict, disorganization, incompetence, or other downright bad stuff going on, I can see that there are principles and theories that outline what *might have* happened instead. Could I have prevented what happened if I knew then what I know now? Probably. But the point of storytelling is not to prevent experience but to learn from it. When I could match these principles with my own experiences, it gave both the principles and my experiences a whole new meaning and worth.   (Scott, Fall 1998)

LINKING SERVICE-LEARNING TO
STAKEHOLDER MANAGEMENT

MG 101 is founded on the vision that all management is stakeholder management (even if stakeholders are poorly treated) and that all people can benefit from a deeper awareness of how to manage. This vision is manifest in three convictions about the place of management in everybody's education: first, that specific knowledge about the activities and skills of managers and of how society's key institutions are structured and governed is an essential component of a general education; second, that well-educated people must have opportunities to develop the skills necessary to coordinate and control cooperative efforts to solve complex problems; and third, that people who will inevitably bear managerial responsibility and exercise formal authority must learn to think broadly and critically about their roles in society (Miller 1991, p. 155). In this sense, MG 101 is consistent with other articulations of the entwined goals of service-learning for social change on the ground and critical awareness in the head. For example, David Lisman (1998) links service-learning to the achievement of civic literacy. Civic literacy is not the study of government structure, but a more profound sense of the self and its role in creating public spaces and goods. In short, since we all live in various and overlapping organizationally bounded life spaces, MG 101 is designed to serve as a liberating experience in organization and management "—liberating in the classical sense: How free can people be who do not know how, or when, or why, or with whom to set and achieve worthy objectives?" (Miller 1996, p. 81). It aims to liberate its (future and present) managers to think critically and act consistent with conscious aspirations.

The key insight of the stakeholder approach is to understand how a company both affects and depends on a much wider set of stakeholders than more narrow theories of the firm, which tend to define away the complexities of multiple stakeholders in favor of explaining firms as merely entities that exist as a nexus of spot contracts or a nexus of lower transaction costs (Freeman 1984). Some approach stakeholder management as a type of sophisticated environmental scan. Although stakeholder theory can be a clearly instrumental practice, Donaldson and Preston (1995) persuasively argued that its fundamental basis is normative. Stakeholders are identified by their legitimate interests in an organization's activity, whether or not the organization has any corresponding operative interest in them. The interests of stakeholders are of intrinsic value. That is, each group of stakeholders merits consideration for its own sake and not merely because of its ability to further the interests of some other group, such as shareowners. Stakeholder management requires, as its key attribute, "simultaneous attention to the legitimate interests of all appropriate stakeholders, both in the establishment of organizational structures and general policies and in case-by-case decision making" (Donaldson and Preston, 1995, p. 67).

Many MG 101 students leave the course with a fairly sophisticated view of stakeholder management—recognizing that their success as individuals and as a collective entails understanding, attending to, and balancing the disparate goals, needs, and contributions of multiple stakeholder groups, including themselves. For example, we offer the following student's reflection:

> I learned more than I could have ever imagined through my involvement in our company's service project. My previous obsession with grades, and my constant evaluation of my self-worth based on these grades, seems absurd to me right now. I am currently taking a course in which I have received consistent C's on exams. If I had received two C's last year, I would have had a nervous breakdown. But this year, spending time with the children at Northumberland Housing Authority has effectively put things into perspective. What I have given and received from the children there is more satisfying and fulfilling than any subjective educational evaluation. For the past year and a half I have been taking classes in order to find a major that conforms to what the majority views as an "acceptable" occupation. Through this experience I have realized that providing the type of service we gave to the children of NHA, and creating these types of relationships, is what fulfills me as an individual, and what I want to pursue as a career.   (Scott, Fall 2000)

Management students, like most people, want to act according to their values. Modern organizations make that acutely difficult (Jackall 1988). Hence, a management curriculum must go beyond making them aware of ethical dimensions because intention to act ethically is not sufficient. By doing stakeholder management with *all* identifiable stakeholders, MG 101 avoids the problems of over-codified and under-practiced knowledge of ideals of ethics or procedural justice. By working with practice and identity, and not just ideas, MG 101 has more potential to affect the deeply held values of our students.

In terms of the design for learning, let's consider one recent company, *Big Time Teez*, a company that sold t-shirts to fund a very ambitious service project; working with a transition program for special needs adolescents (SIMON house), the company built and furnished a library and a hotel room mock-up (for job training). Working with a range of stakeholders led the students to more mediated meanings about their practice of managing. Success for *Big Time Teez* was partially identified by the instructors and the requirements of the class. The negotiation of the meaning of success was further mediated by external and internal stakeholders. The service project's goals emerged from discussions with SIMON house's teachers and students, both of whom the company members considered to be their clients. They designed a get-to-know-you party and a buddy system, matching company members with SIMON house students, so that they could build relationships with, not just learning spaces for, the SIMON house students. Internally, the winning business coalition had pitched a "Tooter's" t-shirt concept as a play on the infamous Hooter's restaurant. Though its proponents argued its irony would undercut the sexism of the product, many in the company were uneasy about this business project. In the middle of the norming phase, with new information about the threats of such a product to their company's image gleaned from talking with Simon house teachers, campus administrators, and potential customers, the whole company changed course and produced a vintage t-shirt that sold well, and that they were proud to give as keepsakes to the SIMON house students.

The *Big Time Teez* story demonstrates that the stakeholders that are relevant for learning in MG 101 are not only external service or business partners. MG 101's impact on students learning the practice of stakeholder managing is in two dimensions: one is the actual service projects; the second, less obvious, dimension is the practice of treating themselves as legitimate stakeholders. Just as stakeholder managing blurs an overdrawn division between external and internal stakeholders, so MG 101 blurs the critical impact on service-learning of "external" clients/partners and "internal" group and individual learning.

THREE PROJECTS

This blurring of external and internal can be seen in the three, interrelated projects each company conducts: business, service, and reporting. The requirement of completing these three projects is identified by the instructors and not open to negotiation. Furthermore, while the *process* of making decisions—deadline dates, group development phases, and so on (see figure 7.1)—for the company is fairly tightly designed, the actual *content* choices—service and business projects (the company's mission), company structure, leadership, and operating plans—remain entirely in the hands of students, leaving much of their practice local, negotiated, and emergent.

The primacy of the service project in MG 101 companies' missions is intentional and long-standing. To provide a broader sense of how the double-bottom-line structure of the companies plays out, we refer to table 7.1, a representative sampling of MG 101 service and business projects. The heart of each MG 101 company's mission is its service project, the objective of

Table 7.1 Sample of representative MG 101 service and business projects

Service projects	Business projects
• **Special Needs:** Converted old home that had been a day care facility into SIMON house—a life skills transition lab for mentally and physically challenged Lewisburg High School students (made entrance, kitchen, & bathroom handicap accessible, painted, etc.); Services for Students, Parents, and Faculty; & donation to 9/11 fund	• BU Novelty T-shirts
	• BU Picture Frames
	• CD of local bands
	• Otis Spunkmeyer™ Cookies
	• BU Swiss Army Knives
	• BU Leisure Pants
• **Special Needs:** Special Olympics Training, Party & Basketball Game for adult Special Olympics athletes	• Gourmet Hot Pretzels
• **Hunger:** Canned Food Drive for the Council of Churches	• BU Frisbee®-style discs
• **Community Development:** Lewisburg & Regional Downtown Survey of Local Business Rental Spaces for Lewisburg Town Council	• BU Folding Stadium Chairs
	• BU Aerial Photo Poster
	• BU Hooded Sweatshirt
	• BU Novelty Playing Cards
• **Fund-Raising:** Walk-a-Thon for Hole In The Wall Gang Camp & Sunflower Day Care Center Cleanup	• BU Insulated Pitchers & Datamatch Dating Services
• **Youth and Poverty:** Northumberland Housing Authority, Children's Services and Thrift Shop Clothing Drive	• BU Fleece Blankets
	• BU Pint Glasses
• **Youth and After School:** Milton YMCA—Basketball Clinic for Regional Kids & help establish basketball league for area children	• Lewisburg Area Shoppers Discount Card
	• BU Nalgene® Bottles
• **Environment:** Shamokin Creek Cleanup	• BU Tank-top Shirts
• **Animals/Poverty:** T&D Cats of the World Animal Shelter— Built Otter Shelter & Union County Food Program— Clothing Drive, built shelves, & organized clothing	• BU Covered Coffee Mugs
	• BU Baseball Shirts
	• BU Polo Shirts
• **Historic Preservation:** Dale/Walker Farmhouse Restoration & education program for children attending Donald Heiter Community Center	• BU Pullover Fleeces
	• BU Calendar
	• BU Hats
• **Mental Health:** Danville State Hospital—Services for the mentally ill	• BU Visors
	• BU Boxer Shorts
• **Seniors:** Riverwoods Senior Citizens Home—Activity Nights & built raised garden & donated to WIC Program	• BU Zippo® Lighters
	• Cow Roulette Event
• **Housing:** Habitat for Humanity construction & Haven Ministry (homeless shelter) evening and overnight work	• Pudding Wrestling Event
	• BU Beach Towels
• **Multigenerational:** PlaySpace of Downtown Lewisburg— renovations and public relations services for new multigenerational learning cooperative	

which is to make as big a difference in the world as the members' creativity, sensitivity, and company resources will permit. Each company's service project must meet important needs that would probably not get met unless the company identified them and then committed its skills and energies to meet them. Each company must identify the needs of various clients it might help fulfill, decide which of those clients' needs it will try to meet effectively, plan and organize efficient methods to meet them, and then implement those methods. This is never a simple, linear process—which is exactly why it is such fertile ground for learning. Students find it much easier to stand in the shoes of their target business customers (mostly students similar to themselves) than those of their target service clients, who usually have social backgrounds and histories quite different from their own. The work they do to investigate, understand, and fulfill (or fail to fulfill) their service clients' needs teaches them powerful lessons about responsible management.

The organizational and managerial challenges involved with working with community partners and service client groups to understand and fill their needs provides much deeper learning about stakeholder management than can be learned by merely organizing and operating a business project. An exemplary case of this is a recent company (*Big Cats, Balls and Bats*) that elected to do a simple, straightforward business project (baseball shirts) in order to meet their own internal passions to do two relatively complex service projects—to build an enclosure for abandoned great cats, and to collect clothing, build shelving, and organize donated clothing for a food and clothing pantry. Students learn firsthand that creating profit alone is not enough to justify an organization's existence. Every organization must strive to contribute in some meaningful way to the society in which it operates, whether it is through providing a needed product or service, creating life enhancing work for the people in the organization, or promoting positive social change. This conviction is more easily communicated when students form and run companies with a dual bottom line—a mission that combines both service and business—than when they organize and operate businesses alone. For a sense of the impact of the novelty of the double-bottom-line approach, consider this student's reflection—echoing comments from many MG 101 alumni—on working with an after school program in a local public housing complex:

> I went into the course thinking I had a fairly firm grasp on what it was about and what to expect. I was completely wrong. Most of the course is undecided in the beginning—what product will be made, what service will be provided, and what every student's part will be in the process of reaching the goals we form together. I underestimated the possibility for the service project to affect me. The most meaningful piece I took from this course was the passion and personal growth that I found through my involvement in the service project. (Liz, Fall 2001)

We have found—through projects serving clients in a regional AIDS resource center, a state prison hospital, and (numerous) senior citizens' centers, for example—the more that students have to "stretch" in terms of working with stakeholders different from themselves, the deeper the learning involved. *Dry Times* was a company that, following the compelling advocacy of one student (who was subsequently elected CEO), undertook a new service partner, the state mental hospital in Danville. The project involved a series of social events to enrich the lives of residents. According to her "CEO's Letter to Stakeholders," working with the mentally ill "raised many questions and doubts in the minds of company members . . . [But] doing a 'new' service project set us apart from other companies, and the challenges we encountered were good learning experiences and brought us together . . . At the end of our final event the question I heard most frequently was, 'So when are you coming back?' " (Abby, Spring 2002).

Moreover, the need to balance the often competing demands of clients with multiple other stakeholder groups (e.g., customers, suppliers, university administrators, MG 101 Board members—professors and teaching assistants, and themselves) prompts students to recognize and explicitly discuss their ethical responsibilities to their multiple constituencies. Among the most dramatic instances of such "prompts" have been projects in which students have "discovered" constituencies whose critical interests they hadn't considered at all. One quick example is a company that, against the advice from many internal stakeholders, decided to sell a t-shirt with a sexually violent message ("Do it Till you're Orange and Blue") to fund a service project at a local community center. As the center became aware of the business product, they nearly withdrew from the project. Such painful discoveries highlight some of the risks of experiential service-learning and designing for communities of practice. At the same time, they are almost universally powerful learning moments for students. We attempt to provide as many prompts for students as possible without sacrificing their own discovery process and before irreversible decisions are made. This is obviously a labor-intensive approach, and one in which the addition of upper-class teaching assistants has helped tremendously. MG 101 is high stakes pedagogy, perhaps nowhere clearer than in managing the fine line between allowing students to make and learn from their own mistakes and protecting them from significant harm.

PRACTICAL THEORY—THREE THEMES

MG 101 companies' achievement of their projects is interwoven with a management curriculum organized around three principal themes, described as a "practical theory of managing" (Miller 1996, p. 78). These three relate to the three functional needs of an organization for effectiveness, efficiency, and community. The triumvirate grounds the reflective and theoretical development the course seeks. Effectiveness, efficiency, and community are the foundation of the practice of management the class expects. These themes show up in the identity of the manager as well. Students literally "work on" several identities in their communities of practice: the manager as efficient, the manager as ends-defining and -achieving (effective), and the manager as invested leader-member (community). All participants experience the juggling act of working on all three identities simultaneously. Consider this example of one student learning about the links between each theme. Here, he explains how discussion of differences among various service projects led to better effectiveness and community.

> Instead of each service coalition bickering and fighting among themselves, they worked collaboratively in order to decide on a mission that the company could effectively, efficiently and enthusiastically support . . . The emergence of inquiry-based decisions not only resulted in what is an obviously more effective way of making decisions, but it also created a stronger community within the company. Instead of personal ownership and emotional attachment to ideas, our company was able to rationally discuss the idea and not the people behind it. In turn, this helped create a community that was more respectful and under-standing of each other's idea and opinions.   (Sophomore Male, Fall 2003)

The practical theory curriculum, along with the predisposition of students to think about management in instrumental terms, makes it common to think of management as only a body of methods. Thinking of management as methods, it is easy to forget that effectiveness, efficiency, and community are also values. There would be no problem if effectiveness, efficiency, and community could always be achieved simultaneously. Managers sometimes find themselves in the happy position where all three values recommend the same decisions and actions. But unfortunately, ethical prescriptions derived from the three values are all too often incongruent. Managers are often faced with choices between politically and/or economically rational alternatives, on the one hand, and humane alternatives on the other. The ethics of effectiveness and efficiency are instrumental, utilitarian ethics—one treats people in certain ways because they are useful in setting and achieving the organization's objectives—whereas the ethic of community is deontological, of human duties—one treats people in certain ways because that's how one is obligated to treat fellow human beings. When these ethics come into con-flict, responsible managers are confronted with a need to distinguish sharply among the three themes, and to decide their priorities. Ultimately, all managers must make hard moral choices. Employing the double-bottom-line standard puts MG 101 students in the position to face those types of hard moral choices. And, MG 101 strives to teach future managers to become intentional in those choices. The interleaving of the three projects and the three themes is one of the key design elements of MG 101 and leaves room for both the reification of meaning and participation in meaning-making.

FINAL THOUGHTS

To begin the long road to changing a management worldview and education wrongly convinced of its power to control people and organizations, we need clear alternatives and approaches to realizing the alternative to status quo management education. Learning stakeholder management by doing it with service-learning is one answer to the need both for alternatives and approaches. The lesson from MG 101 is not to add service to a management curriculum, rather, it is the value of learning management through service.

However, adopting active, collaborative, problem-based, and consequential pedagogies is expensive; it entails commitments of time and energy—physical and emotional—and entrepreneurial risk-taking in the classroom, often in the context of institutional indifference, if not outright institutional hostility. If, as some have argued (e.g., Mintzberg 2004), there is no substitute for full-time (or "real-world," or later) work experience in developing practicing managers, why are we trying to create substitute experiences in the name of better learning and better managing? We believe that service-learning is not a substitute for other experiences; as one former instructor said, "MG 101 is not a dress rehearsal for anything." Rather, MG 101 is its own experience because we know from communities of practice that learning happens as a function of membership in a community of practice, whether designed or not. Inside the world of MG 101 companies, the reality, the consequentiality of *this* place, *these* people, *this* mission, offers a unique opportunity to develop a particular practice of managing as stakeholder managing and a particular identity of the manager as an actor whose decisions are imbued with, and not divorced from, values. To overcome the prevalence of status quo managing, we need something other than simply more experiences. This is a quality, not a quantity, story. We need classes whose design allows for different kinds of experiences. In conclusion, there is a role for designing, but the communities of practice insight reminds us that design is limited in its capacity to control emergent and dynamic communities of practice.

By advocating service-learning and designing for communities of practice, we are not dismissing other experiences, nor are we claiming that the experiences in MG 101 are universally superior. That is to say, the other, "real" experiences of our students are also experiences of practice, but the quality of that practice, the opportunities to negotiate meaning and identity, the control over participation, these are all a function of those other, naturally occurring communities of practice. Coping with extant communities of practice can be very frustrating, humbling, or alienating, not least because orienting and nurturing novices is not valued in those communities. Our commitment, in contrast, is anchored in the learning potential of being in on the ground floor of a new community of practice. The opportunity for service-learning when one is mindful of the communities of practice emerging within and throughout the service-learning classroom and experience is to design more reflection, more openness, and more transparency into the development of practice. Experiential educators have long known that there is a fundamental trade-off between authenticity and learning. If risks are too high, consequences too irreversible, learning can be impinged or hobbled. Simply throwing our students into engagement with practice may err on the side of too much risk, just as asking a novice climber to "just learn" by flailing on an overly difficult climb sacrifices learning for the sake of risk-taking. The challenge for designing communities of practice is to balance meaningful action with opportunities for deeply thoughtful reflection and learning, by creating safe spaces for communicating, providing for flexibility to change course, with greater transparency of processes, and with greater devolution of decision-making.

Finally, MG 101 is high stakes pedagogy for instructors also. There are substantial risks of mislearning, both in terms of content and of the communities of practice within the class. Just as an organization can very efficiently carry out a mission whose ends are immoral, so a service-learning class could contain communities of practice that value and promote the kind of managing we mean to supplant. Herein lie the ethical implications for instructors. How much risk of mislearning is acceptable in the pursuit of consequential learning with authentic stakeholders? There is no simple answer, but we believe that the experience of MG 101 offers insights into this ethical dilemma. The instructors must develop a stakeholder approach to the class itself, and they should welcome the challenges that designing for communities of practice entails.

NOTE

1. We have cited students' lab reports and final papers by first name and date, when available. Some papers were collected under confidentiality agreements and are thus listed only by gender and class year.

REFERENCES

Bornstein, David (2003). *How to change the world: Social entrepreneurs and the power of their ideas.* Oxford: Oxford University Press.

Brown, John Seely & Paul Duguid (2001). Knowledge and organization: A social-practice perspective. *Organization Science.* 12(2): 198–213.

Donaldson, Thomas & Preston, Lee (1995). The stakeholder theory of the corporation: Concepts, evidence, and implications. *Academy of Management Review,* 20(1): 65–91.

Donaldson, Thomas, Patricia Werhane, & Margaret Cording (2002). *Ethical issues in business, a philosophical approach.* 7th Edition. Saddle River, NJ: Prentice Hall.

Freeman, R. E. (1984). *Strategic management: A stakeholder approach.* Englewood Cliffs, NJ: Prentice-Hall.

Godfrey, Paul C. (2000). "A moral argument for service-learning in management education." In Paul C. Godfrey and Edward T. Grasso (Eds.), *Working for the common good: concepts and models for service-learning in management.* Washington, DC: AAHE. 21–42.

Godfrey, Paul C. & Edward T. Grasso (2000). "Introduction." In Paul C. Godfrey & Edward T. Grasso (Eds.), *Working for the common good: Concepts and models for service-learning in management.* Washington, DC: AAHE.

Jackall, Robert (1988). *Moral mazes: The world of corporate managers.* Oxford: Oxford University Press.

Kolb, D. A. (1984). *Experiential learning: Experience as the source of learning and development.* Englewood Cliffs, NJ: Prentice-Hall.

Korten, David C. (2001). *When corporations rule the world.* 2nd Edition. Bloomfield, CT: Kumerian Press.

Lisman, David C. (1998). *Toward a civil society: Civic literacy and service-learning.* Westport, CT: Bergin and Garvey.

Livingston, J. Sterling (1971). The myth of the well-educated manager. *Harvard Business Review,* 49(1): 79–89.

Miller, John A. (1991). Experiencing management: A comprehensive, "hands-on" model for the introductory undergraduate management course. *Journal of Management Education,* 15(2): 151–169.

Miller, John A. (1996). On learning why I became a teacher. In Peter J. Frost & Susan M. Taylor (Eds.), *Rhythms of academic life: Personal accounts of careers in academia.* Thousand Oaks, CA: Sage Publications.

Miller, John A. & Hiller, Tammy Bunn (2004). Management 101: Live introduction to organization and management (44th Edition). Lewisburg, PA: Bucknell University Administrative Services.

Mintzberg, Henry (2004). *Managers not MBA's: A hard look at the soft practice of managing and management development.* San Francisco, CA: Berrett-Koehler.

Samuelson, Judith (2000). Business education for the 21st century. In Paul C. Godfrey & Edward T. Grasso (Eds.), *Working for the common good: Concepts and models for service-learning in management.* Washington, DC: AAHE. 1–21.

Tyson, Laura D'Andrea (2004). "Good works—with a business plan." *Business Week,* May 3, 2004. i388: 32.

Tuckman, Bruce W. (1965). Development sequence in small groups. *Psychological Bulletin,* 63(6): 384–399.

Wenger, Etienne (1998). *Communities of practice: Learning, meaning and identity.* Cambridge, UK: Cambridge University Press.

8

HUMAN RIGHTS–HUMAN WRONGS: MAKING POLITICAL SCIENCE REAL THROUGH SERVICE-LEARNING

Susan Dicklitch

> Where, after all, do universal human rights begin? In small places, close to home—so close and so small that they cannot be seen on any maps of the world. Yet they are the world of the individual person; the neighborhood he lives in; the school or college he attends; the factory, farm, or office where he works. Such are the places where every man, woman, and child seeks equal justice, equal opportunity, equal dignity without discrimination. Unless these rights have meaning there, they have little meaning anywhere. Without concerted citizen action to uphold them close to home, we shall look in vain for progress in the larger world.
>
> Eleanor Roosevelt (United Nations Remarks, 1953)

The perennial problem for the scholar and educator is how to combine scholarly interests with classroom material. Academics have often been castigated for living in the "ivory tower," for being uninterested in the real world, and for being unable to connect with their students. One would think that this would be less so in a dynamic discipline like political science, but like other academics, many political scientists get caught up in covering theories without making those theories real for their students.

Service-learning has become one of the ways academics can link life in the "real world" with life in the ivory tower. Numerous service-learning courses and programs in higher education have developed since the 1990s. According to the National Service-Learning Clearinghouse,

> Service-learning combines service objectives with learning objectives with the intent that the activity change both the recipient and the provider of the service. This is accomplished by combining service tasks with structured opportunities that link the task to self-reflection, self-discovery, and the acquisition and comprehension of values, skills, and knowledge content. (http://www.service learning.org/welcome/SL_is/index.html)

Service-learning by definition, then, should bring the theories we examine as academics to life outside of the classroom. When students ask "why should we study this theory?," academics should be able to show the relationship between theory and the real world. Service-learning courses give students an opportunity to become active participants in their learning experience and engaged citizens in their community. Although important, service-learning is not mere charity or volunteer work. Service-learning is not simply about fixing the broken windows of low-income residents, babysitting for single mothers, or having students file papers for local advocacy organizations. Service-learning is not a college student's charity contribution to the wider community. The key is in combining "service" with "learning." Here the academic role is essential. Service, without a connection to theory or facts (e.g., on poverty, the cause of human rights abuses, or the structure of social service bureaucracies), simply tends to reinforce prejudice and the gap between the students and the "other." Service-learning, when done properly, weds academic rigor with real civic engagement.

Service-learning should force students out of their comfort zones culturally, economically, and socially. It should involve hands-on experience working with others from the community, should challenge students to revisit their prejudices and stereotypes about the "other," and should be able to connect what students learn from theory with reality.

REAL LEARNING

In the spring of 2002, I offered a course in the political science department at Franklin and Marshall College that would change my professional interests and career forever. "Human Rights–Human Wrongs" was structured as a senior seminar that focused on human rights literature in general and U.S. asylum policy in particular. Students were required to apply their newfound knowledge of human rights literature and asylum law to real asylum cases in the York County Prison, a maximum security prison in York County, Pennsylvania that also rents out spaces to the Department of Homeland Security for immigration cases.

Although the course, now in its third year, has changed somewhat, the overall structure has remained the same. For example, in the first two years, the seminar was limited to 16 students, working in teams of two. But, because of enrollment pressures in my department, the course had to be opened to 25 students, so I assigned students to work in teams of three. In addition, the Immigration and Naturalization Service (INS) was disbanded and the Department of Homeland Security (DHS) was created. We now had to deal with two separate agencies that worked on asylum cases: the Bureau of Citizenship and Immigration Services (BCIS) (which handled affirmative application cases) and the Bureau of Immigration and Customs Enforcement (BICE) (which handled defensive asylum applications).[1] And finally, after the second year, I decided to change community partners from a nonprofit immigrant advocacy organization (CIRCLE—Coalition for Immigrant Rights at the Community Level) to a nonprofit legal organization staffed by lawyers (PIRC—Pennsylvania Immigrants' Resource Center) that had the greatest impact on the course.

The student teams worked with asylum seekers from 14 countries: Angola, Cameroon, Cote D'Ivoire, Liberia, Niger, Nigeria, Sierra Leone, Togo, Uganda, Russia, Haiti, Cuba, Jamaica, and El Salvador. In the first two years of the course, our former nonprofit community partner, CIRCLE helped find the cases and in 2004, we worked with the PIRC.

For their academic preparation, students had to read several books, including *Soul of a Citizen* (by Paul Rogat Loeb), *In Our Own Best Interest: How Defending Human Rights Benefits Us All* (by William Schulz), *Detained* (by Michael Welch), and *Do They Hear You When You Cry?* (by Fassiya Kassinja and Layli Miller Bashir).[2] In addition, students were required to read the main human rights documents and the philosophy behind human rights. Seminars were interspersed with guest lectures from community leaders and activists in the field. Students had to apply the readings (like the *United Nations Convention Against Torture and other Cruel, Inhuman or Degrading Treatment or Punishment*, the 1951 *United Nations Convention relating to the Status of Refugees*, and the United States *Immigration and Naturalization Act of 1980*), to their particular asylum seeker's situation. To aid in the technical aspect of the asylum process, students had to read the invaluable *Basic Procedural Manual for Asylum Representation Affirmatively and in Removal/Deportation Proceedings* provided by the Midwest Immigrant and Human Rights Center, and remain in weekly contact with our community partner to ensure that the process was proceeding as smoothly as possible.

In the second year of the course, I asked an alum of the "Human Rights – Human Wrongs" course to work with the students and deal with any technical questions or concerns that they might have. In addition, the student and I put together a "How to" book for the course, which acted as a handbook on what to expect, what to do, and so on. The student received credit from the college in the form of an independent study.[3]

Students, working in their teams of two or three, were assigned a current asylum case on the first day of class. They then had to read through the intake information provided by our community partner (which varied according to how much had already been processed).[4] The teams would work on their asylum case throughout the semester, culminating in an immigration court-ready document (with evidence tables, asylum seeker affidavits) and a legal brief (either in support of asylum or in favor of deportation).[5] This material would be handed over to our community partner who would submit the packet to Immigration Court for the benefit of the asylum seekers. Because the students were not law students, they could not actually represent the asylum seekers in court, but they were required to present their case to the rest of the class and our community partner in a mock trial.

In addition, students were required to keep a daily "reflection journal" documenting their activities as well as their perceptions, experiences, and feelings regarding their experience with service-learning and the asylum process. The reflection journal served as an opportunity for students to reexamine what they learned in their readings and through their real-life experiences with their asylum seeker about human rights and the asylum process in the United States. Many students were stunned to learn that asylum seekers were actually detained in detention facilities and local prisons in the United States. Several of their reflection journals documented that as they became more familiar with the asylum process in the United States and with their asylum seeker's story they became more confident and motivated to commit their time and effort in helping their asylum seeker.[6]

The Process

Most of the asylum seekers we worked with were detained because they had illegally entered the United States (without a visa, or with a forged passport or expired visa). The detainees already had a "credible fear interview" with a DHS official at the port of entry, and the decision on asylum was deferred to an immigration judge with the Executive Office of Immigration Review (EOIR). The asylum seekers were then transported to a DHS detention facility to await a "merits" hearing before an immigration judge. Here, the immigration judge would hear the detainee's story to determine if the detainee had, in fact, suffered past persecution or if he/she had a well-founded fear of future persecution on the basis of the five enumerated grounds for asylum: political opinion, religion, nationality, race, or membership in a particular social group.[7] Human rights conditions within the country of origin as well as the applicant's credibility are taken into account in this hearing. While a BICE attorney represents the U.S. government, most detainees are not in a position to afford a lawyer (nor are they entitled to the equivalent of a public defender). Few speak English as a second language, if at all, and most are completely unfamiliar with U.S. Immigration law.

My students not only interacted with their detainee and learned about another culture firsthand; they also gave a human face to the United States outside of DHS officials, prison guards, and common criminals. More importantly, my students gave the detainees a fair chance at political asylum. Students went to the York County Prison (YCP) to interview their asylum seekers, often through Plexiglas windows, using prison telephones.[8] Through these interviews, they could piece together their asylum seeker's story that would be translated onto the DHS I-589 form (application for withholding from deportation) and the asylum seeker's affidavit.

The interviews were often painstaking processes of trying to communicate with an asylum seeker whose native tongue was not English, and whose culture was totally foreign to most of the students. Once the students got all the details necessary to communicate the asylum seeker's story, they would get to work on finding evidence and case law to support their asylum seeker's story. Students would search multiple documents from Amnesty International, Human Rights Watch, the United States State Department reports on human rights conditions, and case law, focusing specifically on their asylum seeker's story and country of origin. Students would then properly document, tabulate, and present this information in a court-ready legal document. So, if the asylum seeker did not have an attorney at least this file could be submitted as evidence in support of the asylum seeker's case.

The specifics of the cases varied. Some asylum seeker's stories were sad or tragic, but did not meet one of the requirements for asylum (past persecution or a well-founded fear of future persecution based on the five enumerated grounds for asylum outlined in the UN Protocol Relating to the Status of Refugees).[9] These enumerated grounds included persecution based on: political opinion, religion, nationality, race, or membership in a particular social group (including women, homosexuals, or child soldiers).

But, what impact did this have on students? One student opined in her reflection journal, "We're expanding our minds and hearts to be more compassionate citizens; we're learning how to truly be better people, not just memorizing facts and dates and how to apply formulas. I'm learning how to live, not how to make a living. It's refreshing."

Another student working on a case from Sierra Leone, wrote:

> Your class makes students go out into the real world to try to analyze and solve real problems. It also shows us that what we take for granted here in America is not universal and is missing in most of the world. When I communicated with my asylee through the plexi-glass of the jail, I found out how human life can be so drastically different but essentially the same. What I mean is that even though [asylum seeker's name withheld] has experienced more misery and pain than I will ever know he can speak the same language and laugh at the same things that Ryan and I laughed at. He is just a young adult like us yearning to be part of a society that can nurture his amazing potential.[10]

Most telling, one student wrote,

> . . . through service-learning, I have broadened my horizons, learned to think outside the box, and have begun to learn how to be a better person. This class taught me things that I could not have learned through reading a book or listening to a lecture. It has taught me greater compassion and greater acceptance for things and people different from myself. It has taught me how very lucky and fortunate I am to have been given the life and opportunities that I have been given . . . I have learned that individuals can make a difference and that sitting idly watching the world go by is the greatest crime against humanity. I feel my eyes have been opened and a shield has been lifted . . . This course is what liberal education is all about. It would be a grave injustice to the students of this school to be denied such an extraordinary opportunity to truly encounter this valuable and unique learning experience.

Although the students may have felt these strong feelings during and immediately after their service-learning experience, this alone does not prove that this course would have lasting impact on the students or encourage them into civic engagement; student's actions after the class do. One student quit a lucrative computer science job to pursue a Master's degree in human rights theory and practice at the University of Essex. Another became Franklin and Marshall College's post-graduate intern for service-learning, a third completed an internship with the Tahirir Justice Center in Washington, D.C., a fourth will work on an independent study on political asylum reform in the United States, and two others completed a summer internship for credit with CIRCLE. Several others decided to attend law school, with several indicating that they would further pursue human rights or immigration law.[11]

Students did not have to be pushed and prodded to do their work. This is because this course had more than just the traditional course pressures, that is, getting a good grade, being able to answer professor's questions, not letting down the community partner. The students knew that if they did not put in the time to properly research their asylum seeker's story, case law, and find evidence in human rights reports to substantiate the claims, their asylum seeker would most likely get deported. And, if in fact the asylum seeker was telling the truth about his/her human rights abuse, deportation could mean further torture, abuse, or even death. Other human beings, from different cultures, speaking different languages, living completely different realities were depending on my students to make sure that their story got heard. There is no greater motivator than that. Students learned an important lesson in civic engagement: their commitment to their asylum seeker's case made a difference to another human being, even if they were not granted asylum.

REAL SERVICE

But does this success with the students mean that this course provided a meaningful service to the community? At any one time, the U.S. government incarcerates about 22,000 noncitizens in DHS detention facilities and jails, like YCP (Mallone 2001). In fact, the DHS detention facility in YCP is the second largest in the United States. In 2001, 60,853 asylum applications were received, 7,839 were granted, and 14,960 were denied (U.S. Department of Justice 2001). Human Rights First[12] estimates that the U.S. government spends an average of $85 a day to detain a noncitizen. The cost to U.S. taxpayers is $130 million per year. The cost of detaining asylum seekers in particular, is estimated at $42.7 million per year (Human Rights First 2004).

The political asylum process is a huge cost for the U.S. taxpayer, and not a very pleasant experience for the asylum seeker. DHS lawyers are often so overworked that they are flipping through affidavits while they stroll into immigration court. Some immigration judges have become so jaded over the incredible cases that they often don't take the time to read through the asylum seeker's story. Yet, these are not just stories but someone's life experience.

Our community partner, Kathleen Lucas of CIRCLE noted,

> . . . certainly, the research and organization that the students brought to the asylum packets were invaluable to the asylum seekers in their court proceedings. In the cases that have gone to court during and since the class, the evidence was submitted to the judge and was a strong positive factor. There is another more subtle benefit for the asylum seekers that is hard to measure and yet, intuitively, I know it made a huge impact. Asylum seekers in prison usually have no visitors and don't have the opportunity to talk about their cases. They are hesitant to discuss them with other detainees because prison is just not an environment that encourages self-disclosure. Therefore, without the students' involvement, the respondent is telling his or her story to the judge in a fairly raw format. (Lucas 2002)

My students helped win at least six asylum cases, the others (22 cases) were either denied or are still in appeal.[13] Contrary to the perceptions of many, the United States deals very harshly with asylum applicants. Fewer than 20 percent of all asylum applicants win their asylum cases. Only one in five cases are reversed in front of an immigration judge (Midwest Immigrant and Human Rights Center 2002, p. 4). Given these figures, my students did quite well. Students also helped asylum seekers suffering from post-traumatic stress disorder (PTSD). As Lucas suggests,

> Students would ask many clarifying questions, would listen actively and show genuine interest and be supportive of the asylum seeker. Depression and post-traumatic stress disorder are very common among detained asylum seekers and having someone to process the trauma with helps counteract those conditions as well as helping the individual to present their story in a more cohesive manner in court. (Lucas 2002)

It would take time for students to build the necessary level of trust with the asylum seeker to make them feel comfortable enough to retell their story. Our community partner and I guided the students about the importance of being culturally sensitive, patient, and thorough. By the end of the semester, students would feel a very deep bond to their asylum seeker, even if they did not think they had a strong case for asylum. The feelings were mutual. Asylum seekers, often separated from family members and friends, and bewildered by being detained, would have little contact with the outside world. Asylum seekers would send students letters telling them "God Bless You for helping us," and expressing that they were, "overwhelmed with joy to have the students visit them and help them on their cases."

Given the fact that most asylum seekers cannot afford legal help, the students working with our community partner were in many ways the asylum seeker's last hope of having their story documented and supported by evidence. Courses like Human Rights–Human Wrongs may help encourage more lawyers to take on pro bono cases and represent asylum seekers if students do most of the groundwork through research and affidavit writing.

PUTTING IT ALL TOGETHER

Convincing faculty to start a service-learning course may be difficult. The amount of time and effort put into a service-learning course far exceeds that put in to any traditional college-level course. At the same time, the immense satisfaction realized from teaching such a course can never be matched by any traditional college-level course.

There are several key factors that make for a successful service-learning experience for all. One of the most important factors is a strong working relationship between the faculty member and the community partner. There must be mutual respect, trust, and inclusion. Trained in international and comparative politics, I was not an "expert" on the asylum-seeking process in the United States. I had served as an expert witness in asylum cases on Uganda so I had knowledge of how the system worked, but the strengths that I brought to the course focused on the research and theoretical aspect of human rights. I relied heavily on my community partner, especially the first year that I taught the course, to field students' specific asylum questions and to direct students on case law. It was this consulting work (as an expert witness on Uganda) that gave me the greatest personal sense of achievement and involvement in making a real, concrete difference in people's lives. I developed the "Human Rights–Human Wrongs" course in an attempt to give students the same.

But, of course, this is not Hollywood and not everything works as planned. One of the most frustrating things about service-learning is that as the professor you do not always have control over what will happen. For example, I had no control over whether YCP would continue to keep DHS detainees,[14] whether there would be another chicken pox outbreak resulting in another quarantine in the prison; or even if there was a snowstorm that would result in the cancellation of our prison tour. Sometimes the bureaucracy becomes very cumbersome, especially when you are dealing with the DHS. The key to success is to remain flexible (and calm) and to make sure that you have a backup. In the case of the DHS–YCP dispute, our community partner was able to provide a list of "paroled" or "affirmative" asylum seekers with whom the students could work.[15]

Another frustrating aspect of the course was the need to rely heavily on our community partners. In many cases, nonprofit organizations, especially working in immigration, tend to be overworked, underpaid, and largely unavailable when you need them most. Students, often believing that the world revolves around them and their work, became very frustrated in trying to reach our community partner who was understandably pulled in several directions. I realized that I needed to learn much more about the asylum process and become a better resource for my students. I subsequently took a 40-hour basic immigration law-training workshop held by the Mennonite Central Committee (MCC). The PIRC, the nonprofit legal association that was staffed by attorneys that we worked with most recently, offered the students more resources and better access to their detained asylum seekers at YCP.

The downside of working with PIRC was that they also worked with law school immigration clinics and law school students, so our non-law school teams didn't always get the best cases. In fact, many of my students had to work with criminal detainees who were not eligible for asylum, but were seeking withholding from removal.[16]

One of the most successful aspects of the course was bringing in guest speakers, including an immigration judge working for the EOIR, a BICE attorney, and an immigration advocate. Students were able to hear different perspectives on the strengths and weaknesses of the system. The most powerful speaker by far was a Cameroonian doctor who spoke of his torture and persecution in Cameroon because of his attempts to expose corruption in the dispensation of medical drugs.

It is very helpful if you have a college administration and department that supports service-learning. Service-learning is still considered "out there" by many academics. In the third year of teaching the course, I was forced to open my course to 25 students, even though it was set up as a senior seminar. This created a lot more work for me, and changed the dynamics of the classroom. Working in teams of three reduced the teams' burden of work, but decreased the opportunity to have more intimate and lively discussions. I found that I focused more on giving the students the legal background they needed to understand the asylum process rather than discuss the philosophical aspects of human rights and asylum in the United States. However, what was lost in discussion was gained in preparing students for their cases.

This course helped bridge the conceptual separation between human rights at the national and international level. Human rights abuses occurring half way across the globe, came to our backyard in YCP and into my student's consciousness. As educators, it is our responsibility to make learning real and relevant to students, even if in the interim it is more time consuming than we could ever have imagined. However, it does get easier and even more satisfying with time. For a service-learning course to be effective, it has to provide a real service to the community as well as real learning for the student. In an age where America questions its citizen's civic engagement, this course offers hope and a vehicle for students to become more involved and engage in life-long learning and activism.

The true litmus test for successful service-learning is if we have been able to fully integrate the community service into the course, involve our community partners in the planning and teaching, and provide countless opportunities for students to reflect on their learning and service. In this way, students become more than mere tourists in their learning experience; they reap the benefits of their engaged citizenship.

NOTES

1. Affirmative asylum applications are non-adversarial interviews with asylum officers. Defensive asylum applications are held in immigration court, with a government appointed lawyer from the BICE, and an immigration judge presiding

over the hearing. Most asylum seekers do not have formal representation in this adversarial hearing.

2. Kassinja's book was particularly appropriate to this course because it documented the plight of a young woman from Togo who was fleeing female genital mutilation (FGM) and was seeking asylum in the United States. She actually ended up in York County Prison. Shortly after students read the book, they were given a tour of the DHS detention facility in York County Prison.

3. This student was invaluable to the class. If students were unsure about something and didn't want to ask the professor, students were more apt to talk with the student intern. This student wasn't a research assistant, but acted more like a class preceptor.

4. Because we were working with real political asylum cases, we did not have control over what stage of development the cases were in. For example, some students gathered evidence for appeal cases to the Board of Immigration Appeals (BIA) while others were handed cases that did not even have the I-589 (Application for Withholding of Removal) filled out. All the cases provided students with different challenges but similar levels of time commitment.

5. Obviously, if the students wrote a legal brief in favor of deportation based on their evidence and critical assessment of the case the legal brief would not be submitted to Immigration Court in order to avoid any sort of double jeopardy for the asylum seeker. But, it was important to allow the students to critically assess the cases that they were working on, and not to simply assume that all asylum seekers had credible asylum cases.

6. One student wrote, "While it is very exciting to be doing homework that will actually be useful to someone else's life, it is also very overwhelming thinking that if I make a mistake the consequences could be more damaging than just a bad grade on my part." Another student, equally troubled about the burden of working on asylum cases at the beginning of class wrote: "Despite all the stress, anxiety and trouble this class caused, I am glad I took it. I feel I am better off with this experience, including the good and bad thereof than I would be if I had taken a lecture class. I think that's because I was doing something I never thought I'd do, feeling things (stress, relief, and pride) to degrees and in ways I never thought I would in a class, and learned more about me, my work style, stresses, values even though I thought I already knew all those things. They were challenged as never before and that's why the experience, no matter how I felt I succeeded or failed, was worth it." This student worked on an asylum case from Cameroon that was granted asylum.

7. A refugee is defined as "any person who is outside any country of such person's nationality . . . And who is unable or unwilling to return to, and is unable or unwilling to avail himself or herself of the protection of that country because of persecution or a well-founded fear of persecution on account of race, religion, nationality, membership in a particular social group, or political opinion" (8 U.S.C. § 1101 [a] [42] [a]) and UN Protocol Relating to the Status of Refugees, Article 1(2).

8. Students usually have to wait two weeks before they can visit their detainee. At YCP, students must write a letter to their detainee identifying themselves and indicating that they are working with CIRCLE. Detainees will then request the students' names be placed on his/her visitor list. This process can often take a long time. For example, several wings within the INS detention facility had been quarantined for chicken pox. Students, unable to interview the asylum seeker for

his/her story, had to focus on familiarizing themselves as much as possible on the human rights conditions within the asylum seeker's country of origin.

9. Even if the asylum seekers met the criteria of having a well-founded fear of past or future persecution based on one or more of the five enumerated grounds, this did not ensure that they would win asylum. In 1996, the U.S. Congress enacted the Illegal Immigration Reform and Immigrant Responsibility Act (IIRIRA), which introduced statutory bars to asylum including: firm resettlement, a one-year filing deadline, commission of a particularly serious crime, persecution of others, danger to security of United States or terrorism related grounds and commission of a serious nonpolitical crime outside of the United States.

10. This year, the asylum seeker from Sierra Leone won asylum in the United States. One week after he won asylum, I invited him to our class. He forcefully made his case about the relevance of this service-learning course: "Without the student's help, I would not be standing in front of you now, free, telling you my story" (name withheld for confidentiality).

11. Some students became so involved in their asylum seeker's case that they continued to visit the asylum seeker at YCP, even after the course was over.

12. Formerly called the Lawyer's Committee for Human Rights.

13. Even if the other asylum cases didn't result in a granting of asylum my students averaged a 27% grant rate—higher than the rate for most asylum seekers.

14. YCP and the INS were engaged in contract negotiations over the acceptable cost of housing INS detainees in YCP. For the first two months of 2003, the INS pulled about half (400) of the INS detainees out of YCP and moved them to other detention facilities and country prisons in an attempt to make YCP lower the daily costs (and profits) associated with housing INS detainees.

15. The DHS sometimes "paroles" asylum seekers who do not present a flight risk for humanitarian reasons. An affirmative application for asylum occurs when the asylum seeker comes to the United States either legally or illegally and has not been placed in deportation/removal proceedings by the INS.

16. Under U.S. immigration law, if an applicant is able to prove that it is more likely than not that they will suffer persecution on account of the five enumerated grounds, they can apply for relief under the Convention Against Torture, and Withholding of Removal. In some cases, my students were working with asylum applicants that were less than desirable human beings, but they were working within the confines of the law. This in itself, was an important lesson about the legal system.

REFERENCES

Human Rights First (2004). *In liberty's shadow: U.S. detention of asylum seekers in the era of homeland security* (New York: Human Rights First).

Lucas, Kathleen (December 2002). Interview by author, York, PA.

Mallone, Dan, "851 Detained for years in INS centers—Many are pursuing asylum," *Dallas Morning News*, April 1, 2001.

Midwest Immigrant & Human Rights Center (2002). *Basic procedural manual for asylum representation affirmatively and in removal/deportation proceedings.* Chicago: MIHRC.

National Service-Learning Clearinghouse (2001). http://www.servicelearning.org/welcome/SL_is/index.html.

Rogat Loeb, Paul (1999). *Soul of a citizen: Living with conviction in a cynical time.* New York: St. Martin's Griffin.

U.S. Department of Justice (2001). *EOIR, Office of Planning and Analysis, FY 2001 Asylum statistics.* Washington: U.S. Department of Justice.

Vera Institute of Justice (2000). *Testing community supervision for the INS: An evaluation of the appearance assistance program 66,* 1. Vera Institute for Justice.

Walth, Brent (2001). "Asylum seekers greeted with jail," *The Oregonian,* December 10–15, 2001.

9

"No One Has Stepped There Before": Learning About Racism in Our Town

Marilynne Boyle-Baise and Paul Binford

Find something worthwhile in your community that hasn't been researched before. It is hard because no one has stepped there before, but it is rewarding in the end. Hopefully, it will have a high impact on the community.

Y.E., 5/12/04

The Banneker History Project (BHP) used service-learning to reconstruct the history of a once-segregated school in a college town. High school students spent two years gathering information about Banneker School; interviewing elder alumni and probing primary documents for information about its teachers, curriculum, students, and neighborhood. Based on their research, students created an historical exhibit and presented it to the public. The project aimed to raise students' consciousness of local segregation, to foster their social critique, and to help them practice citizen action. It also aimed to engage students in worthy historical investigation and to assist their construction of knowledge from primary sources. In this case study, we describe and analyze the meanings students made from their service work. Students' perceptions offer insights into the potential of service-learning to teach about social justice and to offer an authentic, engaging, pedagogical experience.

The authors were part of a school–university–city–community leadership team for the BHP. Based on her previous work with service-learning in the target neighborhood, Boyle-Baise was invited by local leaders to facilitate the project. Binford joined the team as a doctoral student and research assistant. In this paper, we focus on 25 high school students who served the BHP in its second year. Several questions guide this study: (1) what meanings do students make of racism and segregation from doing service related to social justice? (2) How do students understand history from doing firsthand inquiry into a school's past? (3) How do students think of themselves as citizens based on their participation in this project? Based on responses to these questions, we consider the role of service-learning as a tool for social justice, historic study, and citizenship education. In the next section, we describe the Banneker History Project. Then, we examine literature related to issues of concern for this work.

THE BANNEKER HISTORY PROJECT

The Benjamin Banneker School was named for an historic African American (ca. 1731–1806). Banneker operated as a segregated school for African Americans from 1915 to 1951. In the town, it was commonly called the "Colored School." The school was only a mile from a major, research university, in a seemingly liberal, midwestern town. At its height, the school served 111 students. In the early 1950s, the Colored School closed due to declining enrollment and the repeal of segregation at the state level. The end of segregation was supported a few years later by the famous Supreme Court case of Brown v. Topeka Board of Education. Eventually, the building that housed the school was transformed into the Banneker Community Center that functions to this day.

In 2001, the director of the Banneker Community Center and the president of the local chapter of National Association for the Advancement of Colored People (NAACP) proposed the Banneker History Project as part of an effort by the city to create campus–community partnerships. These local leaders hoped to build interest in the historic value of the school, among youth who grew up around the community center and also among young people city wide. During summer of 2002, a culturally and professionally diverse, school–city–community–university partnership came together to mold the BHP. We followed the original proposal as a grassroots guide for our work.

High school students undertook two tasks. Year one, they collected 18 oral histories from Banneker alumni (also called elders). Year two, they added another oral history to the database. Additionally, they undertook five inquiries related to the school: teacher portraits, census studies, neighborhood mapping, photo searches, and newspaper archives. Students then developed and presented an exhibit of the history they had uncovered.

SERVICE-LEARNING FOR SOCIAL JUSTICE

This project was an effort in service-learning for social justice. Social justice is defined by Warren (1998) as the movement of society toward more equality, support for diversity, economic fairness, nonviolent conflict resolution, and participatory democracy. Adams (1997) and Bell (1997) suggest that social justice education should help students develop a sense of personal agency and social responsibility, analyze oppression at individual, cultural, and institutional levels; and connect social understanding to civic action. Parker (2003) proposes that educators assist students in becoming just individuals who work toward just societies. According to Parker, just individuals should develop a principled conception of justice and a capacity for reversibility or thinking from the perspective of others. Further, they should work toward just societies by interrogating causes for systemic inequities. According to Adams (1997), activism is critical to this approach; without it, awareness can lead to a sense of hopelessness in students.

Wade (2001) describes eight characteristics of service-learning for social justice. It should include: (1) student-centered learning (2) collaborative inquiry, (3) experiential learning, (4) perspective taking, (5) social critique, (6) multicultural emphases, (7) value-based considerations, and (8) social activism focused on equity. Boyle-Baise (2002) calls this approach "multicultural service-learning," or service work that affirms diversity, builds community, and questions inequality. According to Boyle-Baise, multicultural service-learning should offer opportunities for youth to work with and learn from children and adults who live in poverty or struggle with racism, to hear their perspectives, and to glimpse their realities.

This project drew upon a similar effort in social justice education at Renaissance Middle School in Montclair, New Jersey (Anand et al. 2002). A middle school principal and a university professor and parent taught four, nine week courses about a local history of desegregation to a total of 75, ethnically diverse, seventh graders. The students studied firsthand documents and conducted oral histories in order to construct a local story of school desegregation. Doing the oral histories made racism come alive for all students, but students of color reacted differently to accounts of racism than their white peers. Many white students were unaware that racism existed in their town, while some African American students found discrimination painfully familiar. By project's end, students began to discuss discrimination that impacted their own lives, such as having bad clothes or being too fat. Also, doing firsthand research engaged students' learning and taught them useful inquiry skills.

SERVICE-LEARNING FOR HISTORIC UNDERSTANDING

This project put students to work in the service of history. Gallman (2000) suggests that in order for service-learning to enhance historic understanding, students should be helped to make connections between their present, field-based, experiences and past realities. Camarillo (2000) argues similarly. He submits that good history balances historical perspective and contemporary analysis. According to Camarillo, service-learning can provide realistic encounters that prod students to question how current problems are rooted in historic contexts.

den Heyer (2003) submits that students from Western cultures tend to think of powerful leaders, absent from social movements, as sources for change. Similarly, Hallden (1998) found that secondary students personalize history, or think that it is the Great Man or Woman who makes change happen. den Heyer suggests that for social justice projects, like the BHP, teachers should emphasize social movements, rather than individual heroism. He submits that more sophisticated explanations for social change might enhance student's grasp of historic agency.

Joeden-Forgery and Puckett (2000) conducted a local history initiative similar to the BHP. Collegiate history majors helped high school students reconstruct the history of an urban, African American, neighborhood. These researchers focused on perceptions of college, rather than high school, students, but what these students learned informs our effort. College students learned to use methods of historic inquiry, including the use of primary sources, the search for multiple perspectives, and the corroboration of evidence. Through their work, they helped to revive the past accomplishments of blacks in this city.

SERVICE-LEARNING FOR AGENCY/EMPOWERMENT

This project offered students an opportunity to bring a forgotten tale of educational inequality to light. The development of a sense that one can and should bring about change in his/her community might be an effect of this task. Service-learning often is considered a vehicle for civic empowerment, helping youth to develop a sense of agency and social responsibility (e.g., Barber 1992; Battistoni 1997). According to Schwerin (1997), service-learning can empower students by: increasing their political skills, developing their critical consciousness, applying their knowledge and skills, and constructing relationships with community groups.

Some research suggests that service-learning for social justice helps high school students develop political agency and social responsibility. Youniss and Yates (1997) investigated the impact of service-learning on middle-class, urban, black, high school students. Students worked at a soup kitchen, preparing, serving, and cleaning up meals. They took a social justice course in conjunction with their service. Students also learned about homelessness from the kitchen's organizers and diners. In their essays and discussions, students began to develop political agency, to take social responsibility, and to consider differing interests. Students described themselves as willing to "fight for what is right" or to become "a force in the world" (p. 81).

THE INQUIRY

We investigated the BHP as a case, through qualitative, naturalistic, ethnographic study (LeCompte and Preissle 1993). We used an interpretive lens and sought out meanings held by student participants (Atkinson and Hammersley 1994). We looked for multiple perspectives or "voices" and examined potential distinctions in student's race/ethnicity, prior experience with community service, and home neighborhood. We aimed to describe insider's perspectives, to organize categories of meaning, and to seek relationships that might inform service-learning for social justice (Strauss and Corbin 1998).

We took on roles as participant observers; students knew us as leaders and as investigators for the BHP. Our project partners helped us collect interview data and, sometimes, took field notes. On these occasions, they acted as observers-as-participants (Gold, in LeCompte and Preissle 1993). Rules for human subjects were followed strictly: all participants signed informed consent forms prior to the onset of the case study.

Participants

Twenty-five students participated in the BHP. They described their race/ethnicity in the following ways. Three students were male: two were biracial (African American and white and Hispanic and white) and one was European American. Twenty-two students were female: three were African American, two were Indian American, one was Latina (Brazilian American), one was Israeli American, and 14 were European American. There was a mix of juniors (12) and seniors (13). Four students were from the Banneker neighborhood.

Twenty students had prior experience with community service. Most considered themselves involved citizens at the outset of the project, however, meanings for involvement ranged from participation in student council, to occasional, volunteer work at church, or at the community kitchen, to intensive, fund-raising and construction work for the school's chapter of Habitat for Humanity. About half of the students (11) were recruited from Habitat, including five, outspoken, seniors, who emerged as leaders for the project. We also recruited from school clubs for youth of color and from history and government classes.

The students took part voluntarily. They received no course or service-learning credit or grade. We tried to limit the group to 15 to build community and to ease travel for field trips. We set up a simple application process. To our surprise, 25 students applied. We were impressed by the students' interest and intentions, and we accepted all of the applicants.

The Friday Seminar

Very early on, Boyle-Baise and a few local leaders met with principals and teachers from North High School and invited their participation in the BHP. School people thought the Friday Seminar was the best home for the project. On Fridays, an hour was set aside for extracurricular activities, though, occasionally, the time was used for school-wide events or for class meetings. The seminar offered a space to work with students from across the school.

The seminar met about ten times per semester. In addition to class time, we arranged five, three-hour, research-oriented, field trips over the school year. As mentioned earlier, students chose to work on one of five inquiry teams. On some Fridays, inquiry teams did research under the guidance of a project partner. On some Fridays, students met as a group to share and reflect upon their findings. On other Fridays, students read oral history transcripts to get a flavor of the personal impact of Banneker School.

At the end of first semester, some students had lost their overall sense of the BHP. One student articulated this concern well when she said: "we're kind of fumbling and in space" (B.G., 12/12/03). Most students had not participated in the original collection of oral histories, and they felt distanced from Banneker alumni. Some students were frustrated with insufficient progress on their inquiries; they wanted stronger guidelines, clearer foci, and more time for research. Others could not imagine what an historical exhibit of the Banneker School might look like.

Boyle-Baise worked with the students to modify seminar activities for second semester. As per students' requests, the following activities took place. Oral history informants visited seminars to answer students' questions about their inquiries and to retell their school stories. Local leaders who made a difference in town (such as the director of the mayor's office of Safe and Civil Affairs and the president of the local NAACP) visited the seminar to help students think about ways to contribute to their community—beyond the BHP. A university preservationist and a university photographer assisted students in making their exhibit. Students became visibly more engaged in seminar (field notes, 4/2/04), a point supported by their interview comments.

Data Collection

A combination of methods, interview, participant observation, and collection of documents, was used to triangulate data (Denzin 1989). The authors directed the investigation: creating an observation schedule, writing interview questions, and analyzing data. Occasionally, project partners served as "extra hands" and helped to collect interview or observation data. One project partner, a lifelong, near west side resident, and a leader in the black community, served as cultural liaison, arranging interviews with and collecting documents from Banneker alumni.

Project partners helped interview students during seminars in small groups. Three, 30–45 minute interviews were conducted, spaced throughout the school year. In order to gain consistency, scheduled standardized formats were used (Denzin 1989). Field notes were taken during Friday seminars, meetings of project leaders, and field trips. Over 30 hours of field notes were taken. Leaders and students wrote volumes of e-mails related to the BHP. The e-mail communications became another source of data. Students wrote a reflective essay at the end of their experience, which was added to the database. Documents, including data from students' inquiries, were collected.

Data Analysis

We sought to remain open to hearing what students said and how they said it (Strauss and Corbin 1998). We looked for comparative examples in the data, and we examined phenomena through multiple data sources. We also referred to the literature to heighten our sensitivity to meanings that emerged from the data.

Interviews were audiotaped and transcribed. We collated interview responses by question. We used line-by-line analysis and open coding to identify major themes (Strauss and Corbin 1998). We noted names of students next to their ideas and counted the frequency with which ideas were mentioned. We also searched for comparative examples, using constant comparison analysis to deepen our grasp of students' perceptions (Glaser and Strauss 1967). We searched for distinctions related to race, prior community service, and home neighborhood.

We read observation data, first, to get a general description of what happened in the field, and, then, to clarify and corroborate student's self-reports (i.e., interviews and reflective writings). We examined critical incidents noted in field notes to learn more about students' perceptions of the inquiry process, of their own agency, and of racism, segregation, or social justice. We read e-mails, to support trends that emerged from observations and interviews.

RACISM: LEARNING ABOUT BANNEKER SCHOOL

Regardless of their race, ethnicity, or home neighborhood, students were surprised to learn that segregation marked their town's recent past. In this section, we explore students' perceptions about racism, segregation, and social justice. Three themes emerged from the data: "It's all new to me," "racism wasn't so bad," and "it's how people lived everyday."

It's All New to Me

"Our town, once racially segregated? No way! I lived here for 14 years, and this never crossed my mind. Today, our town is very diverse. Just by looking at my high school I can see how true this is" (M.P., 5/7/04). Overwhelmingly, students were shocked to learn that their seemingly liberal, diverse town once housed a segregated school and practiced racial separation. Many had learned about slavery and Jim Crow activities, but considered racism a brutal condition found in southern states. A few students participated in sports activities at the Banneker Community Center, but most students had never heard of Banneker as a community center or of its previous life as a segregated school. "It's all new to me" was a phrase heard repeatedly among white students. These students used phrases like "blast from the past," and "wakes you up" to describe their growing awareness of local racism and segregation. Most students of color, in comparison, said they "learned more" about local racism through the BHP, but found stories of past discrimination to be "things I pretty much expected" (A.B., 10/17/03).

As mentioned earlier, alumni from Banneker School visited several Friday seminars. On one Friday, a professor of African American History, originally from Mississippi, joined them. The guests compared their southern and midwestern experiences with segregation. The exchange became confrontational; the alumni refused to be cast in the southern mold of segregation. Students leaned forward in their seats, listening intently. A few asked questions. As they left the class, we heard remarks like: "awesome," "really real," and "I never heard anything like that before." The next Friday, the same alumni returned to the class. They were quite upset with the professor's view of segregation and wanted to set the record straight. They described the impact of social class bias against local blacks and whites who lived in poverty, and the subtle ways that they knew they were not welcome in stores or restaurants. Subsequently, about one half of the students, both white and of color said they gained new understandings of various forms of segregation and of distinctions between desegregation and integration.

It's How People Lived Everyday

A majority of students told us that hearing personal accounts of segregation (through reading oral histories as well as through speaking directly with elders) was the highlight of the BHP. It made racism personal; something that people lived everyday: "I just think I am learning a lot about individual people instead of the history as a whole. It's been cool to look into the life of one person affected by Banneker School" (R.R., 12/12/03).

Elders told compelling stories about their lives as children in a segregated setting. As examples, one elder described his walk to school. He said he carried rocks to throw at white boys who taunted him and his younger sister. Another elder recalled riding to a children's play at the university auditorium, all dressed up, in a flatbed truck, while white children rode on school buses. As they told such stories, elders became emotional: teary or angry. Students seemed to recognize that racism had a lifelong impact: "elders realized when they got older that they were considered different and, perhaps, even inferior. That's something we forget when we read history, it's much more personal" (L.G., 12/12/03). A third of the students told us they heard views rarely found in their textbooks: "A lot you learn in textbooks is about the South. Hearing personal accounts is something you don't get in history classes. You focus on broader things, but don't hear about local situations unless you do something like this project" (C.B., 5/7/04).

Racism Wasn't So Bad

Elders painted pictures of the near west side as an integrated, communal place: black and white children walked to school together, splitting up at a corner to go to separate schools. Black and white families lived on the same block, and sometimes they worked together to make meals, repair homes, or do crafts. The near west side was small, and a family feeling developed among residents there. Black and white residents lacked material wealth and were ostracized by more well-heeled citizens affiliated with the university.

The railroad tracks created a literal dividing line; the west side of town was the wrong side of the tracks (Field notes, 2/6/04; 2/13/04).

Some white students took these stories to mean that racism "really wasn't that bad." It was "not like completely separate," nor was "everyone out killing and setting houses on fire" (M.H., 10/17/03). These students began to grapple with tacit racism. According to the same student quoted above: "there are different levels of racism which I hadn't realized before" (M.H., 10/17/03). A couple of students tried to figure out why "nobody seemed incensed" (R.R., 12/1/2/03) by racial separation. Maybe, several thought, children don't notice racism. Perhaps, a couple of students thought, it was just a given. These students seemed in the midst of a conceptual shift about racism.

Students of color seemed more attuned to subtle racism. They recognized that there was "a lot of love and friendship in the neighborhood" (J.F., 5/14/04), but no one said racism was "not so bad." Further, some students of color, joined by a budding, white, student activist, struggled to construct a broad, social framework for racism. For example: "I've been aware that certain groups of people, whether by race or class, get the short end of the stick. Now I'm able to point out specific examples and recognize how those occur" (B.G., 12/12/03). These students began to connect their understandings of racism against African Americans to discrimination against other minority groups.

HISTORY: LEARNING THROUGH FIRSTHAND INVESTIGATION

Students thoroughly enjoyed doing firsthand research. Words that described this experience, a first for most, were: "fun," "interesting," and "fascinating." The following categories of meaning emerged from data: "getting to write the story," "what makes history," and "people are still divided up."

Getting to Write the Story

This category was dominant, over half of the students mentioned it. Students seemed intrigued by creating a story, instead of reading it second hand: "all my other experiences with research have been looking it up in a book and the story is right there. But, we're getting to write the story" (E.S., 10/17/03). Students found the research much more interesting than textbook oriented learning: "Getting to know people and listening to their actual stories instead of just reading about what happened helps you understand it more thoroughly" (A.B., 5/7/03). Many students realized that people, especially elders, are important sources of historic information: "Here was someone who lived 70 or 75 years, willing to sit down with someone much younger and expressing that. I realized the special nature a full life can bring" (R.G., 5/12/04).

Students found research to be "very much a process" that took a lot of hard work: "It's not as easy as I thought, but it's really an interesting process.

I love being able to recover old pieces of information and put together a display so that people can learn about the things we've learned about" (G.R., 10/17/03).

Students also realized that historic information was not readily available: "It is kind of hard to find the history. It will be there, but it won't be in good shape at all" (J.F., 10/17/03). Doing firsthand research helped students connect to the story: "It helps you get attached to what you are doing. I can say, I saw this house, I took this picture, I know the guy who lived here" (M.M., 5/7/04). Though challenging, students found the process: "well worth it, it's going to help people in the future see what we've done and realize the history of our town" (M.P., 10/1703).

What Makes History?

As important as getting to write the story, was writing about something noteworthy. Elders made it clear that their experiences of segregation had been either overlooked or misrepresented. Students were given responsibility for a momentous task, that of righting a wrong or telling a long silenced story. This genuine charge seemed to motivate and sustain their hard work of inquiry.

As noted earlier, the main purpose of elders' visits to Friday seminars was to serve as a sounding board for students' ongoing inquiries. During an early visit, in February, a student of color spoke of his group's exploration in the following way: "We are finding out more about the teachers as individuals. We want to honor them and their work. We need to tell a story that has not been told and help people gain respect for that story." He asked the panel of elders: "What message is important to you? What do you wish to be brought out?" (Field notes, 2/13/04). The elders used this forum to respond carefully to his questions. After the class, two white students on the news archive team asked elders for their recollections of "Among Colored People," a separate newspaper column that reported activities of African Americans during the days of Banneker School. One student asked: "Who wrote the column? We have not been able to find out." Another said: "Did you read it? Why were mostly social events reported?" The first student went on: "Mr. Binford told us that we are doing first-hand research. The things we are finding have not come to light before." One elder answered: "It is important to ask us, the people that segregation happened to, about it. We are sensitive to outsiders writing about us without talking to us. You girls bothered to ask. You are doing important work of history" (Field notes, 2/13/04). Students heard this oft repeated message: history should be constructed from the views of people who lived it.

Students learned that history was personal. For the few students who lived in the neighborhood, the Banneker story was a family story. One African American student told us: "A lot of people signed a petition to save Banneker as a center after segregation ended. My grandma signed it and a couple of aunts and uncles too. Then, I grew up at the center" (T.B., 5/16/04). One

European American student told us: "My grandfather and my great uncle grew up in that neighborhood. He knew some of the people who went to Banneker school. They played together as kids" (A.D., 5/7/04). Our data suggest a slight neighborhood effect; students from the area were heavily invested in seeing this story come to light. One student took extra pains to arrange for alumni's visits, another conducted an extra, oral history interview with her grandparents.

Students came to understand that stories of racism can be lost to history; people either do not recognize their worth, or are not proud to tell them. For example, one student told us: "It's very difficult to find the information you want because the community didn't think it was important to keep records of Banneker" (S.L, 10/17/03). Another thought that: "If you look at it from the present standpoint, our town would not be proud of having a segregated school. That kind of history is often lost because people aren't proud of it" (A.B., 10/17/03). Students seemed intrigued by looking for something mistakenly or, perhaps deliberately, cast aside.

One student of color and the budding white activist stood apart from other students. They realized that history is not "single events, all kinds of histories go on at once, there were trends in certain eras" (B.B., 10/17/03). For example, the student of color realized that the presence of a large, research university in town did not save the city from national trends toward segregation. The white, student activist connected jingoist language in old newspaper accounts about Asian people to racism of the day against African American people.

People Are Still Divided Up

Spurred by their historic studies, students wondered how they could confront racism in the present. Project leaders arranged for a panel of "difference makers," local leaders actively engaged in causes for social justice, to speak with students. One panel member, the president of the local NAACP, asked students to consider how often they interacted with groups other than their own. He urged them to: "Think about how many people different from you that you interact with everyday. Get to know someone different from you. Bring down the barriers" (Field notes, C.G., 4/16/04). Subsequent discussion focused on student's awareness of resegregation, particularly of cliques, often racial groupings, that formed in their school.

In final interviews, students continued to tell us that they saw divisions in their school with new eyes: "when I walk through the hallways, I can see that people are divided up. I see little cliques, maybe the African Americans hang out together" (P.D., 5/12/04). Students began to sense the interplay between past events and present realities: "There was segregation in town and there still is. It made me more aware of it now whenever I see it in school" (M.P., 5/7/04). Although there were some divisions, many students "appreciated" their school experiences with diversity. Students said that as they learned about segregation, they gained appreciation for being able to go to school with students from all racial groups. Five students said they appreciated the hardships African Americans endured to gain integrated schools, like their own.

CITIZENSHIP: MAKING A DIFFERENCE

A majority of the students described themselves as involved, rather than as novice, citizens. Many students participated in community service under the guidance of their social studies teacher who was a project partner. Students described their views of citizenship as: "made me feel more like a citizen" or "gave me opportunities to be a better citizen." They explained their aims for civic action as: "set a standard for being involved" or "be an activist."

Made Me Feel More like a Citizen

For students, knowledge of local history made them feel more connected to their town. Over several interviews, about half of the students expressed the following point: "It is making me feel more like a citizen because I am learning about the community I've lived in for 18 years. It is really making me feel more a part of this community" (S.C., 12/12/03). Some described their feelings as finding "place," or as developing "roots." Many students defined informed citizens as better citizens, and found themselves "becoming more aware" of racism in their town. Most students linked their awareness to action: "What is important is to get out there and make people aware of the problems with segregation and stereotypes. Once people are educated about the problem, you can work to create a solution" (S.C., 5/7/04). Quite a few students felt empowered by their work in the BHP. One student of color said she was more comfortable talking about racism. Another felt he had more "control" in representing where he came from.

Gave Me Opportunities to Be a Better Citizen

Over a third of the students already considered themselves as involved citizens. For this group, the BHP offered opportunities to be a better citizen: "It made me aware of things I can do to become a better citizen. I already had this feeling that I should help and make people aware of problems and stuff. It opened my eyes to problems here in this town, not only in the world" (P.D., 5/12/04). For students, like this one, the project "reinforced" or "confirmed" commitments to "be involved," or to "give back to my community." A couple of students said they changed dramatically, from taking a back seat in community service groups, to taking leadership roles. One student explained her shift in this way: "It's easy to be involved in a group, but let other people do the work. I realized if I didn't do the work, no one else would. I learned I can be as active as I want without limitations" (Y.E., 5/12/04).

The project had an ebb and flow. As noted, students expressed dissatisfaction near the end of first semester. Quite a few students, especially those who saw themselves as more involved citizens, felt the project had not yet allowed them to make any real difference: "It's good to be educated and to know what's going on, but I actually want to make a difference in someone's life. I don't think we've made a big impact yet" (C.B., 12/12/03). Later in the project, this problem seemed resolved. For example, one student told us that she was satisfied with the documentation and presentation of the history she had uncovered and she felt the exhibit offered her the opportunity to leave a "permanent mark" on her town (M.M., 5/7/04).

Set a Standard for Being Involved

At the project's end, about half of the group, mostly the white students, described their civic intentions in this way: "Set a standard of being involved, work in the community, and stay informed about what is going on" (C.B., 5/7/04). Students considered the BHP as an opportunity to act as informants for fellow citizens, helping them become aware of past racism. In the future, they planned to be part of the solution instead of part of the problem: "I don't like it when people don't take an action and just stand back in their own little shell. I mean action as in taking an interest in the community they live in, sharing, researching, and doing projects like this" (A.D., 5/4/04). Several students planned to leave their "comfort zone" and "realize I am responsible for my own actions and take the option to make a difference" (G.R., 5/7/04). For some, taking responsibility meant challenging their biases, and then, "making an effort everyday to talk to people of different ethnicities and races" (G.R., 5/7/04).

Be an Activist

There was a racial difference in students' responses about social action. Students of color (and the budding white activist) talked explicitly about taking action for social justice. They viewed the BHP as an activist project: "To me, this is an activist project. I think it is interesting that people are often hesitant to use that word for it" (B.G., 5/12/04). They explicitly discussed racism and planned to be activists. This group used the following phrases to define their social action: "act against injustice," support "anti-racism," and "question oppression." For example, "I plan to be an activist for anti-racism, not only for African Americans, but for other races, or immigrants, or Hispanics" (P.D., 5/12/04). Students saw themselves as "career organizers" or "activists," and several thought they gained skills of leadership and organization that would serve them in good stead in those roles.

Patterns and Possibilities

In this section, we return to the questions that guide this study and consider them in light of the data: (1) what meanings do students make of racism and segregation from doing service related to social justice? (2) How do students understand history from doing firsthand inquiry into a school's past? (3) How do students think of themselves as citizens based on their participation in this project?

Making Meaning of Racism

Students seemed to see shades of racism; to begin to realize that segregation took different forms in different settings and to grasp distinctions between desegregation and integration. In-depth study of one instance of segregated schooling seemed to help them move from surprise about local racism to more complex understandings about how it worked and who it affected. Students also seemed to move from an abstract to a personal grasp of racism. Personal testimonies from survivors of segregation provided direct examples of discrimination that startled the student's assumptions about racism. The strong impact of personal witness to racism suggests that service-learning for social justice should tap into the wisdom of elders and other adults with direct experience of inequality.

White students seemed to gain a good deal of sensitivity about racism, but they still struggled to grasp the insidiousness of tacit racism. Students of color, and one budding, white activist, while initially surprised about local segregation, seemed more ready and able to make linkages to other instances of racism. They also seemed more inclined toward activism focused on inequity. Student's insights and oversights are telling: students might need assistance to check their developing grasp of racism, and some of the students' own peers might provide the assistance they need.

Students discussed racism as "it." This reference suggests that racism is held at arms length. There was some indication that students, especially those of color, gained more comfort in talking about racism, and a majority of students felt the history of racism should be taught. Still, exploration of language use about race seems pertinent to service-learning for social justice.

Understanding History

Although we conceived this project as learning about racism first and as doing inquiry second, the power of investigative learning conditions clearly stand out. Students repeatedly compared abstract, anesthetized information in textbooks with real, personal narratives from elders, and they found textbook-based study wanting. They enjoyed the hunt for information, thinking that they might be the first to discover something long lost. At the same time, they were motivated to dig for something significant, for a story of segregation that had been overlooked. This coupling, of fostering authentic research and of searching for a socially just cause, can not be overstated. Students were pulled in by a hunt for local history, and they were pulled along by a search for a racist blotch on their seemingly liberal town. Students wished for such studies much earlier and more often in their school careers. Obviously, we, educators, need to focus more on what might be called "backyard history." Service-learning can be utilized as a tool to investigate local history, to trace stories of marginalized groups or of discriminatory times.

We encouraged students to make connections between past events and present circumstances, and we provided support (e.g., Make a Difference Panel, and readings from current news magazines about segregation) for them to do so. Apparently, activities impacted students' thinking; they looked at diversity in their high school with new eyes. Yet, we missed chances to discuss significant, real time, and critical moments. As noted earlier, except for field trips, the Friday seminar was limited to an hour. We seemed to tickle reflection just before the class bell rang. Abundant amounts of time are needed to sufficiently reconsider service-learning inquiries, especially when they raise sensitive issues related to social justice.

Thinking as Citizens

The task of reconstructing a history of racism seemed potent for citizenship education: it helped students feel more connected to their city and more empowered as citizens. Students wanted adults to notice what they could do. Their engagement in something as significant as the documentation of local racism was a powerful demonstration of their citizenship. Students saw it as an opportunity to leave a permanent mark on their hometown. It is unlikely that charitable endeavors can satisfy students' aspirations to make noteworthy, local contributions.

Many students saw themselves as involved people who had done community service prior to the BHP. They remained committed during lean times when outcomes of the project seemed hazy. Why? There seemed to be a strong local effect; students responded to a need that hit close to home. There also seemed to be an ardent service ethic; students held steadfastly to making a difference. Again, why? We enjoyed unceasing support from a committed social studies teacher and project partner. From behind the scenes, she fanned the flames of students' commitment. Our findings suggest that service-learning projects should seek to assist local, justice-oriented causes, to include a dynamic teacher motivator, and to recruit student leaders from experienced volunteers.

As noted in the literature (Adams 1997), it seemed quite important to connect student awareness to civic action, otherwise students drifted, uncertain of their purpose. The creation of an historical exhibit stretched student's capacities and commitment to the maximum; they worked nights and weekends to make the display. We worked along with them, stretching our time and talents as well. We submit that, for us and for most students, the excitement and success of the exhibit opening made it all worthwhile. However, a less taxing form of social action certainly might meet the same aims.

There seemed to be a difference related to racial background, or prior experience with issues of justice, in students' future intentions as citizens. A stance toward activism is probably a creature of character, but it also might develop over time, as students experience or acknowledge inequity and connect past learning to new contexts. What it means and what it takes to support budding activists might be a topic worth the consideration of service-learning educators.

ROLES FOR SERVICE-LEARNING

In this section, we return to the literature and consider roles of service-learning as tools for social justice, historic study, and citizenship education.

Social Justice Education

This study supports the role of service-learning as a venue for social justice education. The BHP seemed to help students develop personal agency and social responsibility, analyze oppression, and connect social understandings to civic action (Adams 1997; Bell 1997). It fell a bit short in terms of social analysis; more time for student's reflection on real-life examples of racism was needed. Direct, personal contact with survivors of segregation seemed particularly powerful in the development of reversibility or taking the perspectives of others (Parker 2003). Doing oral histories personalized racism, as in the Anand et al. description (2002). Unlike that effort, both white students and students of color seemed surprised to learn racism existed in their own town, but, like that study, students of color seemed to quickly fit that discovery into their frames of reference about inequity.

The BHP was planned with Wade's (2001) characteristics of service-learning for social justice in mind, and it seemed to actualize those aims. It was an example of multicultural service-learning (Boyle-Baise 2002) in that it affirmed diversity, questioned inequality, and built community. The BHP especially fostered intergenerational relationships and respect.

History Education

Service-learning supported the teaching of history. Efforts were made to link past racism to present discrimination, as suggested by Camarillo (2000). More could have been done with this aim; films and readings on historic, local racism and on current resegregation of schools were under-discussed. Students did not learn about history through the lives of great men and women, as is usual in textbook study (den Heyer 2003; Hallden 1998). Instead, they heard from ordinary men and women who struggled daily with racism. We did not set out to study social movements that impacted segregation, as suggested by den Heyer (2003), but our choice of multigenerational informants responded to this aim. Elders perceived racism differently, depending on their age. These differences spurred discussions of the times of which elders spoke. This study indicates that high school students, like college students, can construct original historic accounts that chronicle the lives and contributions of marginalized groups (Joeden-Forgery and Puckett

2000). This case possibly adds new data to suggest the power of authentic, historic inquiry in pulling students toward deep, prolonged, and complex community service.

Citizenship Education

This study supports the claim that service-learning can civically empower students, helping them to develop personal agency and social responsibility (e.g., Battistoni 1997). Students seemed to develop more critical consciousness of racism, to apply their knowledge in the creation of a public display, and to use their understandings to kindle relationships with other students, local leaders, and elders (Schwerin 1997). As in the Youniss and Yates study (1997), most students seemed to leave the BHP with some commitment to be a civic force in the world.

CONCLUSIONS

The BHP was an exploration in service-learning for social justice that built from "ground zero" as students saw it. We asked students whether such a project should be tackled by others. To a person they said, "yes, definitely." Many students said something like the following: "find something worthwhile in your community that hasn't been researched before. It is hard because no one has stepped there before, but it is rewarding in the end. Hopefully, it will have a high impact on the community" (Y.E., 5/12/04). We hope our portrayal spurs further consideration of service-learning projects such as this one.

REFERENCES

Adams, M. (1997). Pedagogical frameworks for social justice education. In M. Adams, L. A. Bell, & P. Griffin (Eds.), *Teaching for diversity and social justice: A sourcebook* (pp. 30–43). New York: Routledge.

Anand, B., Fine, M., Perkins, T., Surrey, D., & Renaissance School Class of 2000. (2002). *Keeping the struggle alive: Studying desegregation in our town.* New York: Teachers College.

Atkinson, P. & Hammersley, M. (1994). Ethnography and participant observation. In N. K. Denzin & Y. S. Lincoln (Eds.), *Handbook of qualitative research* (pp. 248–261).Thousand Oaks, CA: Sage.

Barber, B. (1992). *An aristocracy of everyone: The politics of education and the future of America.* New York: Oxford University Press.

Battistoni, R. M. (1997). Service-learning and democratic citizenship. *Theory Into Practice,* 36(3): 150–157.

Bell, L. A. (1997). Theoretical foundations for social justice education. In M. Adams, L. A. Bell, & P. Griffin (Eds.), *Teaching for diversity and social justice: A sourcebook* (pp. 1–15). New York: Routledge.

Boyle-Baise, M. (2002). Multicultural service-learning: Educating teachers in diverse communities. New York: Teachers College.

Camarillo, A. (2000). Reflections of a historian on teaching a service-learning course about poverty and homelessness in America. In I. Harkavy & B. Donovan (Eds.), *Connecting past and present: Concepts and models for service-learning in history* (pp. 103–114). Washington, DC: American Association for Higher Education.

den Heyer, K. (2003). Between every "now" and "then": A role for the study of historical agency in history and citizenship education. *Theory and Research in Social Education*, 31(4): 411–434.

Denzin, N. K. (1989). *The research act* (3rd Edition). NY: Prentice Hall.

Gallman, M. (2000). Service-learning and history: Training the methaphorical mind. In I. Harkavy & B. Donovan (Eds.), *Connecting past and present: Concepts and models for service-learning in history* (pp. 61–81). Washington, DC: American Association for Higher Education.

Glaser, B. & Strauss, A. (1967). *The discovery of grounded theory: Strategies for qualitative research*. NY: Aldine.

Hallden, O. (1998). On reasoning in history. In J. Voss & M. Carretero (Eds.), *International review of history education: Vol.2. Learning and reasoning in history* (pp. 272–278). London: Woburn Press.

Joeden-Forgery, E. & Puckett, J. (2000). History as public work. In I. Harkavy & B. Donovan (Eds.), *Connecting Past and Present: Concepts and models for service-learning in history* (pp. 117–137). Washington, DC: American Association for Higher Education.

LeCompte, M. & Preissle, J. (1993). *Ethnography and qualitative design in educational research* (2nd Edition). New York: Academic Press.

Parker, W. (2003). *Teaching democracy: Unity and diversity in public life*. New York: Teachers College.

Schwerin, E. (1997). Service-learning and empowerment. In R. Battistoni & W. Hudson (Eds.), *Experiencing citizenship: Concepts and models for service-learning in political science* (pp. 203–214). Washington, DC: American Association for Higher Education.

Strauss, A. & Corbin, J. (1998). *Basics of qualitative research* (2nd Edition). Thousand Oaks, CA: Sage.

Wade, R. (2001). ". . . And Justice for All" Community service-learning for social justice. *Issue paper: Service-learning*. Denver, CO: Education Commission of the States.

Warren, K. (1998). Educating students for social justice in service-learning. *The Journal of Experiential Education*, 21(3): 134–139.

Youniss, J. & Yates, M. (1997). *Community service and social responsibility in youth*. Chicago: University of Chicago Press.

10

SERVICE-LEARNING AS A SOURCE OF IDENTITY CHANGE IN *BUCKNELL IN NORTHERN IRELAND*

Carl Milofsky and William F. Flack, Jr.

Bucknell in Northern Ireland is a service-learning program that places students in the midst of a struggle over national identities, the sectarian conflict involving many Catholics and Protestants. It is a three-week, May-term experience comprising two courses. In one, students participate in community-based organizations in L'Derry (the city is "Derry" to the Catholics and "Londonderry" to the Protestants; "L'Derry" is a current compromise) either doing service-learning or working on research projects, some of which become the first phase of honors theses. The other course provides lectures by academic and social leaders representing diverse experiences of, and points of view about, the conflict. They tell about development of the civil rights movement and relate their personal experiences by referencing the American civil rights movement, the feminist movements, personal experiences engaging social class deprivation, and struggles of national liberation around the world. Intensive experiences in community organizations, working with people involved in social and political change, teaches both students and faculty that, in Northern Ireland, politics, civic action, and personal lives are inseparable.

One way in which we might understand the challenge to participants in our program is summarized by the New Left slogan from the 1960s, "the personal is political" (Aronowitz 1996). On an isolated, elite liberal arts campus like Bucknell University, it is easy to see politics as distant, and people affected by serious social problems as "the other." We are absorbed in the social and academic routines of a community that emphasizes careers, credentials, and consumption, where the experience is intensely personal but politics are often distant. As service-learning has grown, several programs have moved in a direction like the Northern Ireland program, where community projects are increasingly merged with academic teaching, and where the service elements work with local people to address issues of social and political injustice. One example that predates our own program is the

Bucknell Brigade. This is a program in Nicaragua where Bucknell staff and students first went to provide disaster assistance in the wake of Hurricane Mitch, and have continued donating resources and labor to build community infrastructure, like a new health clinic. Students have also become involved in women's employment issues in Nicaragua. Other programs involve students in local community issues and projects.

While the term "service-learning" implies a curriculum in which staff teach civic values to students, programs such as those in Nicaragua and Northern Ireland are as significant for staff members as they are for students. As for our students, the comfortable separation between private and academic lives on one side, and troubled social conditions on the other, are disrupted. Both staff and students are challenged to think about our own ethnic identities, our religious orientations, and our willingness to consider ourselves members of the same world as those who struggle with personal issues linked to economics and politics. As professionals, we are challenged to make our expert knowledge available and useful to people who may neither understand nor care about our intellectual, disciplinary boundaries. We must find ways for our teaching, research, and service work to be meaningful to our field hosts while at the same time not violating conflict of interest policies set by our administration.

Bucknell in Northern Ireland is an intensive program, both intellectually and emotionally. It involves full immersion in the culture of Northern Irish society for three weeks. Despite the relative peace that has endured in the wake of the Belfast/Good Friday Agreement, sectarian division remains intense; what was once a violent conflict has now largely shaded over into a psychological one (Cairns and Darby 1998). Students learn details about life in a small society dominated by conflict. Some of the field placements lead to quick intimacies, in which students learn how residents of L'Derry have experienced killings, open street battles, and state and paramilitary violence. Not all students share the experience of developing intense personal relationships with L'Derry residents, but for many of those who do, the experience can be profound. It becomes more than the sort of vivid instructional experience where we engage history with immediacy. Students' identities are challenged.

One way in which identities are challenged is in terms of social and political consciousness. Personal life, social or community life, and political processes are deeply intertwined for many of the residents of L'Derry, because everyone has been affected by the sectarian conflict. Nearly everyone has a friend or relative who has been killed or badly injured, and all have had their daily lives disrupted by bombings, street fights, and police or paramilitary blockades (Morrissey and Smyth 2002). Many of our hosts believe passionately that each individual has a responsibility to understand the political situation, and to take public action to ameliorate the situation (Deane 2004). The better we have come to know people in Northern Ireland, the more we feel encouraged, and at times directly challenged, to become clearer about our own perspectives on the conflict, and about our own political identities.

At a programmatic level, we also have been challenged to understand the political significance of our time in L'Derry. Since many outside the country still believe that Northern Ireland is a dangerous place, few outsiders take the time to get to know the community, and to learn about the many dimensions of the troubles that have disrupted Northern Irish society for the last 35 years. Outsiders' willingness to understand these experiences is important for a small country of less than two million people in the far northwest corner of the European continent (a favorite saying is that "the next parish is Boston"). As Americans, we look at the local conflict with outsiders' eyes. We are allowed to hear diverse and conflicting points of view, some of which members of one faction or the other have themselves not yet heard. As naïve but interested observers, we are also allowed to ask questions that residents themselves may not ask for fear of ostracism or retribution. These questions sometimes come up in the personal conversations students have in field settings. It also happens in formal presentations, such as the panels of community leaders that we organize to give students opportunities to ask questions about local and national progress on efforts at reconciliation and peace-building.

A DESCRIPTION OF THE PROGRAM

In this section of the chapter, we expand on the brief description of our program given in the opening paragraphs. *Bucknell in Northern Ireland* began as a collaborative effort between a sociologist (Milofsky) and a clinical psychologist (Flack), with shared interests in personal and interpersonal dynamics, and in field/internship-based learning. We started with a research interest, and began working with colleagues at the Centre for Voluntary Action Studies and the Centre for the Study of Conflict on the Coleraine Campus of the University of Ulster. We shortly became interested in taking students to Northern Ireland, and traveled with a similar, short-term program organized by Professor Jon Van Til, a sociologist at Rutgers University. With Van Til, we developed a consortium for education and research in Northern Ireland that includes Bucknell, Rutgers, Swarthmore College, and the University of Ulster. Although the consortium has not yet played a significant role in our local program planning, it has provided us with a collegial group from which we obtain intellectual and emotional support.

Bucknell in Northern Ireland is a short-term, study-abroad program, based at Magee College, the L'Derry campus of the University of Ulster. It takes place during the three weeks between the end of our spring semester and the start of our summer session. Our choice of L'Derry as the home base for this program was based on the seminal role of the city and its inhabitants in the history of the "Troubles"[1] (vernacular for the 30 or so most violent years of the sectarian conflict in Northern Ireland, roughly 1968 to 1998; see Fraser 2000, and Mulholland 2002, for recent, concise summaries), on its relative peace and stability during the past decade, on its small size and lively atmosphere, and on the availability of numerous nonprofit organizations (the "voluntary sector," as it is called in the U.K. [Williamson, Duncan, and Halfpenny 2000, and Birrell and Williamson 2001]) as service-learning settings for our students. The timing of the program is also important, because students who, for a variety of reasons, cannot study abroad for an entire semester have an opportunity to do so during a brief (albeit intense) period of time. The relationship we have forged with colleagues at Magee College is crucial, especially because they handle most of the logistics of arranging lectures by faculty and community leaders, and provide contacts with community organizations.

Academically, the program includes two courses. One is a core seminar covering the history of the island and the country; integral topics such as the Troubles, parading and paramilitaries, and the peace process; and related topics such as psychological trauma and the role of the voluntary sector in societal reconciliation. Students learn from their readings, lectures, guest speakers, site visits, and panel discussions that, for a tiny country of less than two million people, Northern Ireland is a very complex society, with local and national politics seemingly far more complicated than their own. Although initially overwhelming in its complexity, most of our students are surprised to find that they have developed a sophisticated understanding of the causes, characteristics, and consequences of the sectarian conflict by the end of their time in the program.

The other course is a service-learning experience, in which students are matched with nonprofit organizations in L'Derry, based on their backgrounds, academic majors, and interests. Students are required to spend two full days each week in their organizations, with the twin goals of learning about the history, current work, and future prospects of the organization, and of providing some meaningful service to the organization.

We encourage students to consider carefully the links between what they learn in the seminar, and what they learn in the community organizations. In fact, as the program has developed, we find that the distinction between academic and community-based learning has become increasingly blurred. For example, starting in 2003, we have organized a moderated discussion panel, consisting of four local community leaders (two each from the Protestant/Unionist/Loyalist tradition, and from the Catholic/Nationalist/Republican one), focused on consideration of the ideas raised in a recent government document (*A Shared Future: Improving Relations in Northern Ireland*, 2003) about extant problems underlying polarization between the two communities. The basic idea is to bring together representatives of both communities, in order to communicate their positions on key issues to the students (and to each other), and to clarify areas of agreement and disagreement on those issues. Students are required to fashion specific questions based on the rather ambiguous material included in *A Shared Future*, and the session is recorded.[2] Students write up the panelists' responses to the issues, and provide a detailed analysis based on the entirety of what they have learned throughout the program; this constitutes their final examination in the seminar.

Field Experiences Challenge Identity

Bucknell in Northern Ireland (a) challenges the often comfortable and complacent identity zone in which students live on campus; (b) shows them that private lives can be intensely and surprisingly intertwined with politics; (c) involves them with people and organizations many of which take surprisingly courageous actions to build civil society; and (d) allows students to bring the disciplinary training of their major to bear on research problems. Students grow emotionally, and they also come to understand differently the often abstract, decontextualized knowledge that they have learned on campus. This section offers examples of these challenges at work.

People often describe the Bucknell campus life as a "bubble" because it is rural, because many of the students come from backgrounds of economic privilege, and because the student body conveys a sense of homogeneity. As instructors, we see more diversity than some students recognize among themselves. However, the class and residential origins of our students, coupled with an intense social world, discourages many (i.e., white, upper-middle-class students) from recognizing potential points of tension in their identities.

The Northern Ireland program seems to change this state of affairs for some students, along two dimensions. One involves students' personal identities and histories. Gender and ethnic identities are key elements in the cases we describe. The other dimension is the link between analytic inquiry and those matters in which students are emotionally invested.

Discovering Personal Identity

For many students at Bucknell, issues of ethnicity, gender, and religion are not prominent topics of discussion. Although we frequently notice that they tend to self-segregate by race, for example, when eating meals or choosing living quarters, we hear little concern expressed by students about these kinds of matters. Similarly, although we hear from many students that their social lives are largely controlled by the fraternity system, we see little evidence of large-scale concern about gender dynamics on campus. The pressure to conform to a single standard for identity, to become a "Bucknellian," seems to be both powerful and widespread. In addition, despite resources such as a Women's Resource Center, and annual programs such as V-Day, university-sponsored programs for the dissemination of information about, and personal support for, issues relating to gender, ethnicity, and religion are frequently targets of public criticism from politically conservative students. Given this context, Arlene (not her real name[3]) was surprised to find that in Northern Ireland, she came to care a lot about her ethnicity, gender, and religion.

Case Study

Having grown up in a small town, in a family where her father was a stay-at-home dad and her mother a professional, Arlene is a strong student who came to Bucknell with intense professional goals. She began as a premedical student, then changed to sociology, and arrived in Northern Ireland as a person with considerable field experience in service-learning and ethnographic learning courses. She was ready to accept a challenging field experience and wanted to learn about the enduring causes of conflict.

Without reflecting much on her Irish Catholic heritage, we placed her in a Protestant Loyalist community organization, supervised by an articulate community leader, William Johnson, whose primary employment was as director of an organization for individuals who formerly had been imprisoned as members of Protestant paramilitary organizations. During the course of the placement, Mr. Johnson was continually available to Arlene, gave her personal support, made sure she always felt safe, included her in his family life, arranged experiences in other Protestant community organizations, and brought her into activity planning and implementation meetings for his organization.

He was extremely pleased to have an intern, since few students are placed with Protestant community organizations.[4] His members benefited from hearing how a relatively uninformed outsider heard and reacted to their points of view and their experiences. They also appreciated that with a Catholic, they could show their tolerance toward individuals whom one might expect to represent the opposition. Arlene did useful work for them, but a large part of the service she provided came simply from her presence. The success she and her hosts had in developing a safe and cooperative experience, and the experience of "otherness" that her presence represented, were extremely important.

Despite the safety and practical success of the placement, the two important things Arlene learned were that she was Catholic, and that she was a woman. On some level, of course, both of these were obvious to her when she arrived in L'Derry, but in another sense, she had lived assuming that the world was largely an egalitarian place. In the United States, her ethnicity did not matter much, and did not seem to have much to do with her sense of self. As an ambitious, preprofessional woman, she did not think that her gender really mattered much in learning and at work.

She was shocked in the organization, however, to find members talking casually about how killing Catholics was not a bad thing. The conversations were not a matter of casual bigotry. Her hosts viewed members of the Irish Republican Army (IRA) as active and dangerous adversaries. They pointed out that the only people who had been killed in recent years in their neighborhood were Protestants. Our students placed in Catholic areas certainly heard equally bitter words from their field hosts. Furthermore, these conversations did not preclude simultaneous discussions of cooperative projects that the community group ran with Catholic organizations, and with particular respected individuals on the Catholic side.

What shocked Arlene was simply that the people her hosts talked about killing were people of her own religious group. While not feeling personally at risk, she suddenly felt a kind of allegiance and solidarity that she had never felt in the United States. She had to confront the fact that she was different, and that she took her difference from others as a potentially serious and unpleasant reality about herself. Although buffered by her nationality, Arlene came to share emotionally in this problem to a degree that she had not foreseen.

On the gender front, Arlene was similarly treated with courtesy and respect. She was included openly in activities, and men in the organization fully explained their work and their ideas to her. However, Northern Irish society is a gender-divided society. Arlene could not ignore the fact that in most of the settings she saw, there were no other women. Furthermore, the ethos and style of the groups were gruffly male in terms of the language, humor, and high level of alcohol consumption that were the norm. Her gender difference was sharply driven home when Arlene and her Bucknell field partner were invited over to Mr. Johnson's house for dinner. Although the men prepared dinner, it was clear that Arlene was to spend the evening with Mr. Johnson's wife and talk about women's issues like what American convenience foods, like Kool Aid, she particularly liked. The reality of gender segregation created a tension in her field placement that made Arlene's gender identity an issue in a way it never before had been in her life.

Upon returning to the United States, Arlene spent the summer at home, intensely reflecting on the identity challenges she experienced. She felt especially challenged on the gender front. However, returning to campus, she also found the "ideology" of homogeneity that permeates Bucknell simply external to her life. She had focused her career interest, having decided to enter nursing, and embarked on new field experiences that allowed her to work with nurse midwives serving pregnant teens. Where previously she had been an anxious student, she now seemed relaxed and confident. Placed in field settings where field supervisors were not ready to give her the kind of placement experience she wanted, Arlene worked creatively and assertively to create the setting she wanted to combine helping others with developing her own professional service ideals.

INTERTWINING THE PERSONAL
AND THE POLITICAL

For many Bucknell students, American society is massive, monolithic, and comfortable. While there is a sprinkling of political activism, most students would describe the Bucknell campus and its students as politically disengaged. Among the most surprising discoveries about Northern Ireland is that such a small, seemingly homogenous society, made up of people who are typically friendly and engaging, can be so fraught with chronic, complex interpersonal conflict. What also makes the experience vivid for many students is that their field hosts in Northern Ireland are often intensely interested in them as people. They spend a great deal of time talking to them and exchanging personal life histories. In the context of telling personal stories of horror, loss, challenge, and bravery, they often ask students about their values, and about how they would respond to the difficult situations that have been commonplace in L'Derry.

Case Study

Greta was placed in one of the L'Derry organizations that seeks to build peace by encouraging residents to tell their personal histories in group settings. A coalition of organizations had developed in the city following a roughly similar methodology. Many residents had experienced extreme trauma, and had coped with injury, conflict, and loss by blocking the events from their memories. All that remained was psychological depression and deep hatred for the opposite side. Greta was placed with Tom O'Hearn, a clinical psychologist who was one of two paid staff members in a non-denominational, faith-based initiative.

Beginning with a new client, O'Hearn would first encourage individuals to talk about their personal histories in a private therapy session. Once they were able to talk about their experiences, he would have them tell the story again and again in various group settings of others from his sectarian group. Typically, as people retold their stories, they would remember more details, and be better able to express their sense of fear, outrage, and anger. As their narrative became more familiar, it also became less emotionally evocative to them and somewhat routinized. This then made it possible for them to tell their story in a group that included sympathetic members of the opposite sectarian group. Ultimately, the goal was for individuals to tell their stories in mixed groups that had not been pre-screened for "safe" members of the opposition. Two feature films about Bloody Sunday that dramatized this storytelling process were premiered in L'Derry at this time.

Greta was assigned to work in the organizational office with O'Hearn as her supervisor, and he provided her with the usual menu of office chores and tours to different facility settings where his group worked. From the beginning, however, O'Hearn spent most of Greta's field days talking to her, and they developed the habit of taking long walks around town. He told her of his experiences growing up in the Bogside (a working-class Catholic section of L'Derry), of his friends in the IRA, of his personal experience of the Bloody Sunday disaster, and of the years he spent out of the country (having left to escape the conflict). He also asked her incisive questions about her own life, her personal history, and her family. More importantly, as they walked he presented her with difficult and challenging questions about values' choices, and pushed her to think about how she would respond to situations he had encountered. He showed her that she harbored hidden prejudices and talked about lessons he'd learned that forced him to erase those feelings and attitudes from his mind.

As field supervisors, we were somewhat perplexed by the intensely emotional relationship that developed between Greta and O'Hearn. Yet there did not seem anything inappropriate about the relationship; he took Greta to meet his wife, and they too had intense conversations. Then, as we heard other students' stories, we discovered that many of them had also experienced conversations where Northern Irish hosts pushed them to examine deeply held convictions, habits, and beliefs.

We came to see that many people in Northern Ireland do not seem to relate to people as members of categories, but rather relate to them in more unique and individual ways. This leads to intensive, reflective conversations. Furthermore, since the sectarian conflict has so deeply affected the lives of most people in Northern Ireland, many have developed an abiding sense of how personal choices interface with political realities and actions. They certainly were as interested in the laissez-faire style of engagement with political realities that our students revealed, as our students were fascinated by the zero-sum-style of politics (Arthur 2000) shown by their hosts.

COURAGEOUS CIVIC ACTION

Many service-learning placements in the United States are with organizations that either function as routine service providers or engage in noncontroversial helping work. Students in Northern Ireland, by contrast, are often placed in organizations where volunteers courageously and creatively engage in conflict, and work to defuse potentially explosive situations. Students hear that the purpose of volunteering is to challenge one's limits, and to take risks that might solve problems believed by some to be too difficult to tackle.

Case Study

Genevieve is an African American woman who stood out as dramatically unusual in the starkly Caucasian, homogeneous society of Northern Ireland. She was a student who joined our program from another college, a somewhat introverted chemistry major, who wanted to follow up a long-term interest in peace studies. We arranged a placement with a women's peace organization in which Genevieve and another student would take oral histories of women who had played a significant role in the peace process in the L'Derry area. The women's organization is a loose coalition of women involved in various kinds of work, but who collectively had played a profound role in peace making. They were part of the women's coalition that had elected parliamentary representatives and facilitated negotiations leading to the Belfast/Good Friday Agreement. In L'Derry, they had been leaders of the storytelling movement described in the last case, taking the lead themselves by telling painful stories about personal losses. They also had organized an initiative to counter rumors often believed responsible for outbreaks of conflict. Protestant and Catholic women worked together, using their cell phones to call each other from opposite sides of community dividing lines, to check on the accuracy of stories that were provoking anger and violence. In one year, this action was believed to have reduced the incidence of violence by 75 percent. Genevieve and her fellow student collected stories in the hope of producing a local history of women's actions.

While the stories of civic action were moving and compelling for both students, the more powerful impact came simply from hearing about sectarian prejudice and violence. Their familiarity with racial tensions in the United

States did not seem to have prepared them for what they saw and heard about beatings, death, prejudice, and violence in Northern Ireland. Genevieve commented that she simply felt "as though she had been hit in the stomach." She felt nauseated hearing about the extreme violence attested to by her respondents.

As we talked about the interviews and her reactions, Genevieve talked about gaining a different view of racism and prejudice. Being in Northern Ireland objectified the process for her. Her feelings of shock arose from seeing how she had internalized racism in the United States so that it had become part of her identity. Observing a similar conflict so closely in another society, she found herself stepping away from the role of combatant and this required intensive self-examination. It also led her to see how deeply opposed she was to conflict. When she returned to campus, she changed her major to political science. She currently intends to do graduate work in peace studies.

FIELD RESEARCH AS DISCIPLINARY STUDY

From the outset, one of our goals for *Bucknell in Northern Ireland* has been the development of ongoing, collaborative research with our Northern Irish colleagues, in which students might play a meaningful role. To this end, we have encouraged students to consider helping with projects of ours (e.g., a collaborative study between Flack and Professor Ed Cairns at the Coleraine campus of the University of Ulster, aimed at identifying verbal and nonverbal sources of conflict during dyadic interactions between Catholic and Protestant college students), or to develop projects of their own. In the last of our case examples, we briefly describe the work of a student who chose to conduct a pilot project using the program itself as a source of data.

Case Study

Myra is a student interested in pursuing an academic career as a clinical research psychologist. At the time of her participation in the program, she had become especially interested in the emotional aspects of service-learning, a topic that she would subsequently develop into an honors thesis. Myra's position as the program teaching assistant, coupled with the development of positive relationships with her peers in the program, made it possible for her to collect data from her fellow students on their emotional responses to the program. With approval from Bucknell's Institutional Review Board, she asked the students to provide both quantitative and qualitative information about their emotional responses prior to, during, and after the end of the program. Myra collected information based on standardized rating scales of emotional states, as well as detailed descriptions of students' emotional expectations, experiences, and outcomes based on personal interviews.

Although her sample was not large enough to warrant statistical analysis of her quantitative data, Myra did uncover some interesting trends. In general, students appeared to start with largely positive, pleasant emotional expectations, to move through a period of difficult, unpleasant emotional reactions to the combined stressors of heavy academic expectations coupled with intensive experiences in community organizations and precious little "down" time, to a final resolution in which their responses were a complex mixture of both positive and negative emotions. Changes in students' emotional responses seemed to mirror their increasingly sophisticated understanding of the academic information obtained in the seminar with the emotional learning gained in the field. Although disappointed in the limitations of her pilot data, Myra became convinced of the importance of emotional aspects of the program, and subsequently went on to develop a more comprehensive study of the role of emotional responses to service-learning experiences across departmental boundaries at Bucknell.

ENGAGING IDENTITY

Earlier we mentioned the 1960s slogan, "the personal is political" and these cases demonstrate that, for some of the students, *Bucknell in Northern Ireland* encouraged them to see how their private identities were fused with wider social and political realities. We have used the language of identities because identity is an extremely important theme and issue in Northern Ireland. While there are important economic, political, historical, and geographic aspects of the sectarian conflict, being Protestant or Catholic suffuses residents' lives, defining where they live, where their children attend school, and even where they will shop and, in some cases, which side of the street they will walk on. If the personal is political, then new social movements driven by people oriented toward a particular identity issue become the way we do politics (Aronowitz 1996; Skocpol 2003).

One way of understanding the troubles in Northern Ireland is as an identity movement that, over the past 30 years, has gradually erased the multiple political commitments that informed the early years of the Northern Irish civil rights movement.[5] Students are challenged most directly on the level of their own identities as the cases illustrate. Our discussion of changes in identity includes both the personal level of discourse covered in our examples, and what C. Wright Mills (1959) called creation of a "sociological imagination." Mills did not see this type of imagination as the property of sociologists. Citizens, journalists, or poets could have more vivid sociological imaginations than sociologists. We believe that our students gain a sociological imagination through their experiences in Northern Ireland.

Mills argued that Americans too often understand their experiences in a narrowly individualistic way. We all have personal troubles that preoccupy us and, to different degrees, oppress us. Political power and domination are maintained, he argued, because we see our problems as solely personal, and thus see social trends and the arena of political action as distant and separate from the personal, emotional realm. Things change dramatically when we come to recognize that many others experience and confront personal troubles that are the same as our own. We may come to see that we share a situation that is the product of particular historical developments, or the result of a political context where we are put into a difficult situation with many other people. We may also come to see that we suffer from economic deprivations that others share, and that are created by the actions of powerful individuals or groups.

When we recognize that our personal troubles are not just ours alone, but that they are products of broad-ranging societal processes or structured patterns of inequality, our views may change. We no longer have to blame our own personal failings for our difficulties. We can see possibilities in joining with others to take action. We also can see new possibilities for living as individuals. For example, when disabled people see their situation as partly the result of political decisions that create barriers to independence, they may come to see new possibilities for autonomy and competence, and become less handicapped and dependent (Acheson 2001). In this sense, Mills argued that the development of a sociological imagination could change lives.

On one hand, the personal troubles students learn about in Northern Ireland are products of community battles, army incursions, and funded government programs. On the other hand, the students learn that inactivity and disengagement are virtually impossible in Northern Ireland. It is not a matter of joining a side and fighting, although some students do affiliate with political groups around current issues (e.g., participating in political marches promoting the scheduling of national elections in June 2003). Rather, individual citizens know that civic actions they may take can matter. Often those civic actions are privately challenging and painful. However, if one does not learn to accept private pain, the public problems may continue and worsen. Our students are challenged to recognize that withdrawal into a comfortable private sphere, whether in Northern Ireland or in the United States, may not be practical because it allows social and political oppression to go unchallenged. For some, their identities change because they come to see their private, personal selves in sociological terms.

CONCLUSION

In our experience at Bucknell, service-learning programs generally make a significant contribution to learning by taking students off campus, allowing them to interact with local residents whose social class backgrounds contrast with the background of most students. Engaging in service involves students in substantive professional or community development issues that strongly encourage them to recognize and develop their own personal competence, while learning contextual information about settings that builds knowledge.

Service-learning in the local community has an important, but somewhat limited impact because students return to the campus and its social and cultural life at the end of the day. They do not tend to become fully immersed in the culture of the local community, and so they may not grasp the differences between that culture and the culture of their own, campus community, to which their identities remain securely anchored.[6] The organizations where students are placed also tend to be consensus oriented, offering professional services or community building. Political organizations in inner city areas may challenge students to reexamine their values, but in Bucknell's rural community there are few placements that encourage students to examine the role politics plays in their personal lives, or that cause them to see how personal life, civic action, and political participation and effectiveness overlap.

Service-learning in the *Bucknell in Northern Ireland* program is different. Community placements often challenge students' personal identities, their understanding of how the political dimension intrudes into private life, and their understanding that volunteer work can be creative, courageous, and significant in the context of social conflict. Our service-learning placements in L'Derry have all been safe and personally supportive, and our hosts have been excited to have intense, even if brief, experiences with American students. These placements often challenge students in ways that elicit difficult emotional responses, and require their integration with an increasingly sophisticated, intellectual grasp of the issues on the ground. Previously comfortable assumptions are thrown into doubt, and new thinking about the self and about the nature of social life, is often the result.

This chapter draws on a perspective that sees its ultimate goal as helping students to develop toward increasing cognitive and emotional complexity. When this happens, students become less dependent on external group norms and values in making important personal choices, and increasingly empowered to use sophisticated thinking to become more autonomous. Students move through a series of developmental stages as they proceed through college, and their development is motivated by emotionally challenging experiences that cause them to reconsider less mature habits and assumptions.

A service-learning program like *Bucknell in Northern Ireland* provides experiences that challenge students, so that they must examine entrenched assumptions about themselves and respond. These challenges are focused on students' assumptions about personal identity, about the importance of involvement in community volunteer work, and about how the political realm is relevant to intimate facets of life and related decisions. Within the framework of our service-learning program overseas, it is difficult for students to escape the deep personal contradictions that their experiences inevitably present. At the same time, the structured reflective writing and personal support our program provides, coupled with our sympathetic and supportive field hosts, allow students to do this work in an environment that feels safe and caring. We see significant personal, intellectual, and emotional growth as a consequence.

NOTES

We gratefully acknowledge the comments made by Deirdre M. O'Connor, Director, Bucknell Writing Center, on earlier drafts of this chapter.

1. Probably the best-known event of this period is "Bloody Sunday," which took place in the Creggan and Bogside residential districts of L'Derry on January 30, 1972. Thirteen Catholic men were killed by British paratroopers during a civil rights march protesting the government's policy of interning Catholics without trial (see McCann 2000 and Mullan 1998).

2. Our hope has been to share with our hosts in Northern Ireland extensive video recordings we have made of lectures and interviews with community people. Currently, we are exploring the possibility of creating an Internet site that would allow people to download video, under a partnership between Bucknell and the University of Ulster.

3. The four case examples in this section are based on our observations of individual students who have participated in *Bucknell in Northern Ireland*. We have changed their names, but not their genders or races, and have received permission from each of them to present their stories in this chapter.

4. Given the emphasis in both popular and scholarly accounts on the Catholic civil rights movement, and the nationalist/republican political thinking often associated with this movement, it is not surprising that many students tend to empathize more readily with the Catholic/nationalist/republican side than with the Protestant/unionist/loyalist side. For a recent scholarly account of current issues from the Protestant/unionist perspective, see Porter (2003).

5. Lectures and panels are an important part of the *Bucknell in Northern Ireland* program. One of the most successful of these was a panel of early civil rights activists Bernadette Devlin-McAliskey, Eamonn McCann, and Ivan Cooper, chaired by L'Derry community organizer Eamonn Deane. They told us that in the 1960s, the Bogside, widely seen as a homogeneous Catholic ghetto in L'Derry, actually had a substantial number of Protestant residents. The first demonstrations in the civil rights movement included many Protestant participants, partly because it was a student movement and the students came together across sectarian boundaries. The movement became an identity movement over time because UK politicians, the media, and the government (controlled by Protestants) emphasized religious differences rather than class conflicts or political injustice as the major cause of the fighting. One of the main effects of the 30 years of conflict is that Northern Ireland has become dramatically more segregated and divided (Northern Ireland Statistics & Research Agency, 2001). Thus, there is some justification in saying that the conflict in Northern Ireland is primarily an identity conflict.

6. The same dynamic occurs in many semester-long study-abroad programs, in which students often report that they affiliate exclusively with other U.S. students, and thus do not develop close, meaningful relationships with members of their host society.

REFERENCES

Acheson, N. (2001). Service delivery and civic engagement: Disability organizations in Northern Ireland. *Voluntas*, 12: 279–294.

Aronowitz, S. (1996). *The death and rebirth of American radicalism*. New York: Routledge.

Arthur, P. (2000). *Special relationships: Britain, Ireland and the Northern Ireland problem*. Belfast: The Blackstaff Press.

Birrell, D. & Williamson, A. (2001). Voluntary–community sector and political development in Northern Ireland, since 1972. *Voluntas*, 12: 205–220.

Cairns, E. & Darby, J. (1998). The conflict in Northern Ireland: Causes, consequences, and controls. *American Psychologist*, 53: 754–760.

Deane, E. (2004). An intentional community. *Fingerpost* (Spring Edition 2004): 107–108.

Fraser, T. G. (2000). *Ireland in conflict 1922–1998*. London: Routledge.

McCann, E. (2000). *Bloody Sunday in Derry: What really happened*. Dingle, Co. Kerry, Ireland: Brandon.

Mills, C. W. (1959). *The sociological imagination*. New York: Oxford University Press.

Morrissey, M. & Smyth, M. (2002). *Northern Ireland after the Good Friday agreement: Victims, grievance and blame*. London: Pluto Press.

Mulholland, M. (2002). *The longest war: Northern Ireland's troubled history*. Oxford: Oxford University Press.

Mullan, D. (1998). *Eyewitness Bloody Sunday*. Dublin: Wolfhound Press.

Northern Ireland Office (2003). *A shared future: Improving relations in Northern Ireland*. Belfast: Office of the First Minister and Deputy First Minister.

Northern Ireland Statistics and Research Agency (2001). *Northern Ireland census of population*. Retrieved October 4, 2004, from http://www.nisra.gov.uk/census/start.html.

Porter, N. (2003). *The elusive quest: Reconciliation in Northern Ireland*. Belfast: The Blackstaff Press.

Skocpol, T. (2003). *Diminished democracy*. Norman, OK: University of Oklahoma Press.

Williamson, A., Duncan, S., & Halfpenny, P. (2000). Rebuilding civil society in Northern Ireland: The community and voluntary sector's contribution to the European Union Peace and Reconciliation District Partnership Programme. *Policy and Politics*, 28: 49–66.

11

Service-Learning as Crucible: Reflections on Immersion, Context, Power, and Transformation

Lori Pompa

I was caught completely off-guard by how powerful Inside-Out has turned out to be. This program has been a learning experience unlike any other that I have had the privilege to be a part of, and it has absolutely had an impact upon the way that I view the world.

Mike, outside student

I didn't expect to learn so much. I didn't expect to grow and change as a result of the process. . . . As I reflect on the power of this course, I am awestruck and humbled . . .

Catherine, outside student

Each of these students spent three months in jail. Week after week, they attended class with other Temple University students and a group of individuals incarcerated at the facility. Having class inside a prison is compelling—an experience that's hard to shake. And that is one reason we do it. I don't want my students to shake these encounters easily; in fact, I want the students to be shaken *by* them. I want them to analyze what they experience and question it all: who is locked up and why, how these decisions are made, what these institutions are all about, and what each of us can do to change the situation.

Total Immersion: Learning Content through Context

The prospect of sitting in a classroom with a group of "prisoners" for the first time in my life was an obvious curve ball . . . especially after considering that just a week prior to this experience, I was told that more than half of my new peers were serving time for murder. Therefore, when the guards slid closed the metal gates behind me, I admit that I was nervous.

Gene, outside student

The gate slams shut, the key turns in the lock, and suddenly, we are in a world that is no longer comfortable or predictable. This kind of learning changes lives: it disturbs where we are comfortable, challenges what we thought we knew. This is "Inside-Out" or, by its more formal title, "The Inside-Out Prison Exchange Program: Exploring Issues of Crime and Justice Behind the Walls." And this is how it began.

Sometime in the mid-1990s, I took a class on a tour of a state prison three hours away from Temple's campus in Philadelphia. At the end of the tour, the students and I met with a panel of life-sentenced men incarcerated there for a group discussion. We began discussing with the "lifers" issues of economics, politics, race and class, and—related to it all—crime and how our society responds to it. One of the men on the panel remarked how beneficial it would be to have an ongoing dialogue about these and other issues throughout the semester. Everyone agreed, while realizing that the distance was prohibitive. However, the seed was sown—and grew into a course that Temple has offered now for eight years at two different sites: the Philadelphia Industrial Correctional Center (PICC), a county facility just 25 minutes from campus, and the State Correctional Institution at Graterford, an hour away. To date, approximately 900 students (from the "inside" and the "outside") have taken part.

> Most college courses are lectures and readings which, later on, we are supposed to apply to real-life situations. This class was a real-life situation itself. The readings gave all of us facts, statistics, and the opinions of the "experts," but the class itself was what gave the course an additional meaning and another dimension. The students in the class gave it life—we taught each other more than can be read in a book.   (Kerry, outside student)

Each semester, I take a group of about fifteen Temple students to one of these prisons for class. In the fall, we meet with men; in the spring, with women in PICC; and in the summer, with men in Graterford. We hold class once a week for two and a half hours and address a separate topic each session, including such issues as: what prisons are for; why people get involved in crime; the myths and realities of prison life; victims and victimization; and the distinction between punishment and rehabilitation. Except for an initial briefing and final debriefing, the entire course is conducted in prison, bringing together "outside" students and "inside" students, within the setting that serves as part of the context of the learning. The outside students and I are provided a unique window into the vicissitudes of the criminal justice system, and the more we go in and out each week, the deeper and more complex the questions become. It leads to a process of exploration for everyone involved: we read and write, listen and discuss—but primarily, we experience and reflect upon the experience.

> I have learned so much about so many different issues, including everyday life. This class has been more than "just a class." It was [a] process of getting to know and understand the issues that so many [people] have to deal with every

day of their lives. It was also a process of getting to know myself and realizing who and what I am. (Angela, outside student)

As a particular model in the service-learning genre, Inside-Out affords college students an experience of immersion, providing direct exposure to the exigencies of the particular context of prison, while engendering deep interaction and connection with the men and women incarcerated there. It is the ultimate border-crossing experience. In taking class together as equals, borders disintegrate and barriers recede. What emerges is the possibility of considering the subject matter from a new context—that of those living within that context. The interplay of content and context provides a provocative juncture that takes the educational process to a deeper level.

> . . . [W]e . . . discuss issues within the assigned texts, but we do so by talking to the core of those issues. In learning the perspective of the individuals these issues affect, a greater understanding of such things develops. A conversation is worth a thousand textbooks. In learning only facts, it is easy to forget how they impact on the lives of real people, which is quite significant. (Diane, outside student)

As Parker Palmer (1993) put it: "We do not learn best by memorizing facts about the subject. Because reality is communal, we learn best by interacting with it" (p. xvii). This unique educational experience offers dimensions of learning that are difficult to achieve in a traditional classroom. Inside-Out provides a life-altering experience that allows outside students to contextualize and rethink what they have learned in the classroom, gaining insights that will help them to better pursue the work of creating a more effective, humane, and restorative criminal justice system. At the same time, the program challenges men and women on the inside to place their life experiences in a larger social context, rekindles their intellectual self-confidence and interest in further education, and encourages them to recognize their capacity as agents of change—in their own lives as well as in the broader community.

> It is one thing to discuss issues of criminal justice with other prisoners, but the expansion of one's ideas, beliefs, and concepts are better challenged and stimulated by those who are outside of the process. (Tyrone, inside student)

However, much more occurs in the exchange—layers of understanding that defy prediction. In our discussions, countless life lessons and realizations surface about how we as human beings operate in the world, beyond the myths and stereotypes that imprison us all.

> Inside-Out should come with a warning label—in big black and yellow letters: **Warning: May cause severe damage if taken internally**. We have seen, first hand, the kind of damage the program can do to preconceived notions, stereotypes, and most importantly—ignorance. Even though we came from

Temple with different motivations, some of us curious tourists and sightseers, some with deeper reasons, we also came here incarcerated, mentally incarcerated, and we have learned things about not just . . . the Prison Industrial Complex but about ourselves. Inside-Out has acted, for many of us, as a kind of eye-exam for the soul, forcing us to realize what we believe and why we believe it. And we now realize that our vision was never 20/20. We leave here with a little better vision.   (Glenn, outside student)

A FRAGILE CALCULUS: POWER, PERCEPTION, AND PATRONIZATION

Too often community service is structured as a one-way activity in which those who have resources make decisions about the needs of those who lack resources. It is one more example of the "haves" of our society shaping the lives of the "have nots". . . . Service ought to be a two-way relationship in which all parties give and receive . . .

Rhoads 1997, p. 127

I must admit that I have never been comfortable with the phrase "service-learning." Unless facilitated with great care and consciousness, "service" can unwittingly become an exercise in patronization. In a society replete with hierarchical structures and patriarchal philosophies, the potential danger of service-learning is for it to become the very thing that it eschews. And it can happen in subtle ways.

The crux of the problem revolves around the issue of power. If I "do for" you, "serve" you, "give to" you—that creates a connection in which I have the resources, the abilities, the power, and you are on the receiving end. It can be—while benign in intent—ironically disempowering of the receiver, granting further power to the giver. Without meaning to, this process replicates the "have-have not" paradigm that underlies so many of our social problems.

. . . I realized that Inside-Out was not about who was the best writer, or who was the smartest. I found it was about what we could teach each other. (Maalik, inside student)

One of the difficulties relates to limited or faulty perceptions of this relationship. For example, if we think in terms of assets and liabilities, it is easy to assume that many of the settings that we utilize for service-learning would be considered in the "liability" category. These settings often are places where there are people with needs, a reality that can become the primary filter through which the setting or group of people is viewed. It is a question of definition, perception, judgment—not unrelated to the process of labeling. There is a difference, for example, between "a homeless man" and a man who is homeless, or between "a prisoner" and a person who is in prison. Labeling or perceiving in such a limited way skews the person's identity, resulting in a relationship with the *liability*, rather than with the *person*.

The opening exercises allowed each person to get a glimpse into the other's humanity. Labels such as "inmate" and "student" fell away and were irrelevant. We were just people engaging each other on a basic human level. (Tyrone, inside student)

Mutuality: Moving from "Doing For" to "Being With"

Through the other, we come to experience the self. Mutuality is about how we both give and receive because we connect to the other through a concern, which, in the name of caring, bridges whatever differences we have.

Rhoads 1997, p. 139

We speak often of the resources of our institutions of higher learning and our responsibility to make those resources available to the wider community. While this notion is certainly true, it can be dangerously one-dimensional. What we may fail to see in the equation are the many ways in which the academy *needs* the assets of the community—those tangible and intangible gifts that challenge, deepen, and enhance the world of higher education. The danger is that, without mutuality, service mimics charity, and ". . . charity does not encourage the intimate connections and the personal relationships that result from service built on mutuality" (Rhoads 1997, p. 128).

I now have a better perception of not only some major issues involving the criminal justice system, but these issues have now taken on a personal meaning. . . . I was unprepared for the emotional identification and passion I now feel about the issues discussed in Inside-Out. (Mike, outside student)

At its core, service-learning is about relationship—"a relationship that is based on equality and collaboration. . . . From such a perspective, . . . service is seen more as an act of working with people in need rather than working to serve them" (Rhoads 1997, p. 8). The concept of relationship implies a connection, an interchange, a reciprocity between people. Everyone involved in a service-learning encounter—community members, students, instructor—is impacted upon by the others and by the shared experience itself.

The whole experience has acted as a lens, bringing that part of the world into greater focus. . . . It has left me feeling separated and connected. My life is so different from the experience the [incarcerated] students endure that it puts up a wall of ignorance, but at the same time, learning does the hard job of pulling each brick away as the connections strengthen between not knowing and knowing, and between human beings. (Diane, outside student)

Part of the power of service-learning comes from the dialogic interaction that takes place between and among those involved. This dialogue occurs on many levels and is multidimensional in character. It certainly includes the spoken word, through which participants share ideas, perceptions, perspectives, analyses, critiques—verbalizing realities with and for one another. Fundamentally, it fosters an atmosphere in which people feel increasingly free to "speak their lives," encouraged by the simple yet profound act of being together—in reciprocity, dignity, and gradually developing trust.

> Through the . . . women [in prison], I also learned things about life, not just about prison. Seeing the strength of these women who are living a nightmare made me stronger. I found role models in a very unlikely place.   (Kim, outside student)

The approach to service-learning used in Inside-Out provides a reciprocal arrangement—everyone serves, everyone is served. One group is not "teaching" the other; rather, we are all learning together. In fact, the two groups quickly become one, through a series of community-building exercises. The "service," therefore, is less a question of "doing for" than "being with," in a mutual exchange. In this way, if anything is "done for" the inside students, it is being afforded value as human beings with ideas and experiences to contribute, an opportunity that is extremely rare behind bars.

> From the outside, [the prison] looks enormous. The first time I arrived, I felt tiny and powerless in the presence of so much wall and bars. As I continued to return, [the prison] both shrank and grew. Outside, it did not look as huge. However, on the inside, it became monolithic, cramped with the lives of individuals kept from meeting their potentials.   (Diane, outside student)

A growing number of the incarcerated men and women who have been involved in Inside-Out have become interested in returning to school, an idea often abandoned many years before. For individuals who frequently have been seen—and sometimes see themselves—as nothing more than a liability in life, the class offers an experience of being considered an asset, often making a significant difference in how they then envision themselves.

DEMYTHOLOGIZATION: DEFYING INITIAL ASSUMPTIONS

At the beginning of any college course, everyone involved—whether consciously or not—carries assumptions about those with whom they will be sharing the semester. These judgments can be based on many factors: age, skin color, accent, dress, where one sits, how one acts—whatever is picked up through visual cues. In the Inside-Out class, the usual assumptions are expanded to include perceptions of such things as intelligence, dangerousness, open-mindedness, and trustworthiness. This is the result of two seemingly disparate groups coming together in one space, with the underlying presumption that their respective worlds and sets of experiences are radically different. It takes only a short time for this premise to be dispelled. We are left with one group, whose common elements emerge more prominently than their differences.

All I could see when I sat in class was their "blues" and I think that caused me to subconsciously form false perceptions. However, this all began to change for me when I . . . began to look at the [incarcerated] students as individuals, not just as blue uniforms. . . . This was a very different perception from my first and was an enlightenment for me as an individual. I don't think I have ever felt such a strong change occur inside of me and it will be something that I hold inside for the rest of my life.   (Gina, outside student)

As Robert Rhoads describes in *Community Service and Higher Learning: Explorations of the Caring Self* (1997), ". . . overcoming our sense of alienation involves recognizing real differences, and, at the same time, understanding that we can build some common connections—that the stranger is not so different from myself and that we can engage one another in a common struggle or cause" (p. 119). In the context of the prison, through direct interaction with men and women who are usually marginalized and mythologized, the myths are debunked, and the true complexity of the social justice/criminal justice nexus can be ascertained.

As hard as it was for me to break down my barriers, each of the women penetrated me with inspiration and knowledge and for that I will be forever grateful. Before I started this journey to see beyond our similar colors—their blue prison clothes and my blue uniform—they were women with no names or faces, because all I'd see was the crime not the person. Having been given the opportunity to release my feelings of opposition, through our collective efforts and intellectual exchanges, I can honestly say I'm beginning to shed the "cop" in me.   (Gladys, outside student)

Drawing Forth: Reconstituting the Role of Teacher

I must take responsibility for my mediator role, for the way my mode of teaching exerts a slow but steady formulative pressure on my students' sense of self and world. I teach more than a body of knowledge or a set of skills. I teach a mode of relationship between the knower and the known, a way of being in the world. That way . . . will remain with my students long after the facts have faded from their minds.

Palmer 1993, p. 30

I see my role as facilitating a learning process, by creating an atmosphere in which those involved can experience, examine, and explore. "A primary responsibility of educators is that they . . . recognize in the concrete what surroundings are conducive to having experiences that lead to growth" (Dewey 1997, p. 40). The key is to provide compelling situations, because they do just that—compel us to go further, deeper, in an attempt to understand more fully. This perspective takes the focus off the instructor as receptacle and dispenser of knowledge, challenging learners to take increasing responsibility for their own education.

> My brain never stopped processing information as each student was able to add
> a piece to the steadily growing mosaic. For me, this is what a college class is all
> about. I left class with my mind racing to place all of the pieces discussed into
> their proper places.   (Stan, inside student)

Through a participatory methodology, theoretical knowledge is enhanced
and deepened in ways that are difficult to replicate within the context of a
solely didactic pedagogy. If we conceive of the process of education as
"drawing forth," as its etymology suggests, we can then understand these
contextualized experiences as conduits through which newly integrated
realizations can emerge.

> Despite the classes I have taken, I only know theory not the reality. I had to
> bring myself out of my shell. It is much easier in a classroom at a university to
> be shy. No one puts you on the spot. In many ways, everyone is in their own
> little world. Inside-Out scrambled that up. The whole point was to be in the
> same world.   (Diane, outside student)

In *A Pedagogy for Liberation* (1987), coauthored by Ira Shor and Paulo
Freire, Shor mentions an example in which ". . . the professor learn[s] along
with the students, not knowing in advance what would result, but inventing
knowledge during the class, with the students. This is a complex moment of
study. . . . The material of study is transformed. The relationship between
the professor and the students is re-created" (p. 86). There are many challenges in this approach, not the least of which relates to power and control.
This pedagogy calls for creating an atmosphere in which there is room for
the unexpected to emerge, as well as space for power to be shared among all
participants.

At the inception of the Inside-Out Program eight years ago, my primary
focus was on the Temple students and what they would learn through this
ongoing encounter. Initially, the syllabus called for the students to go to the
prison every other week, and meet on campus during the alternate weeks.
My assumption was that we would need to debrief each session as we went
along. After the first week in the prison, however, the students strongly
recommended that we go to the prison every week. They were concerned
that, in processing separately from the rest of the group, the class would
remain divided, compromising the integrity and mutuality of the exchange.
Their intuition was exactly on target, and gradually led me to a deeper
understanding of the contours of this exchange.

Elsewhere in *A Pedagogy for Liberation*, Freire (1987) discusses the "directive" role of the teacher, in this way: that the teacher is ". . . *not* directive of the
students, but directive of the *process*. . . . As director of the process, the liberating teacher is not doing something *to* the students but *with* the students"
(p. 46). This description is reminiscent of the power of what Parker Palmer
(1993) calls a "learning space." He suggests that this "space" emerges with
a teacher ". . . who not only speaks but listens, who not only gives answers
but asks questions and welcomes our insights, who provides information and

theories that do not close doors but open new ones, who encourages students to help each other learn" (p. 70).

> I look forward to this class. I find it inspiring. There have been points, deep in our discussions, that I've found myself feeling at home. I know that sounds odd, but it's quite true. It must be from a combination of things. I think we all feel generally respected, affirmed, and supported in our opinions and world views. And I feel in this dialogue of Temple students and men on the inside an extremely critical engagement with issues of suffering and our society's account-ability to the widespread phenomenon of suffering.   (Anisa, outside student)

The heart of this methodology is in providing a framework within which the issues that we are studying can be examined in depth. This exploration is mediated through an ongoing group process, in which everyone is afforded the space to raise questions, challenge each other, offer diverse perspectives, and wrestle with the idiosyncratic nature of our system of crime and justice. My hope is that, by the end of the semester, each participant has developed more than merely the ability to take in information, but rather, the capacity to inquire, analyze, critique, challenge—and be challenged by—the information acquired.

> As much as we say we are open-minded, it is not until we are forced to listen to the opinions of others that we really can appreciate the perspective that each of us brings to a subject. This was clearly instructive for me personally. (Tyrone, inside student)

I want my students—those on the outside and those on the inside—to know the issues thoroughly, especially as they impact on their own lives and that of others, and to then take an active role in addressing issues of crime, justice, and incarceration as they are played out in the public arena.

CONTEXT: KNOWING MINE, RESPECTING YOURS, CREATING OURS

> . . . We cannot learn deeply and well until a community of learning is created in the classroom.
>
> Palmer 1993, p. xvi

In shifting the focus from the passive acquisition of knowledge to a fully inte-grated, dynamic process of discovery, an essential ingredient is dialogue. As its name suggests, The Inside-Out Prison Exchange Program is a course of study through which an exchange takes place involving both "inside" and "outside" students. The group generally numbers between 30 and 35, evenly divided between college students and men or women in the prison. This exchange offers an opportunity to delve into topics of concern in the criminal justice arena in great depth. Through both small group interaction and large group discussion, issues are grappled with in a constructive, dialogic fashion.

> Through shared dialogue, a multi-leveled educational process begins that
> awakens a spirit of activism for social change. Participants come to recognize
> universal commonalities and connections among all people . . .   (Paul, inside
> student)

The dialogic nature of the interaction has more to do with a deep exchange,
". . . a moment where humans meet to reflect on their reality as they make
and remake it" (Shor and Freire 1987, p. 98). The dialogue is more a lived
reality than merely a spoken one.

> Inside-Out has changed me so much; it honestly showed me what life is about.
> In the eight years that I've been incarcerated, I've never felt so strong about
> wanting to make a change.   (Maalik, inside student)

An issue that we explore during this guideline process is the idea of context—
understanding that we each have one—and that our unique context, and all
that has helped to form it, impacts how we hear, speak, and take in our
surroundings. Wrestling with complex issues in which varying perspectives
emerge calls for all participants to extend themselves and suspend their judgments in order to maximize the learning for the group as a whole.

> Even when opinions differed, it was striking to note that sometimes I thought
> both were right or equally reasonable. I had to redefine my concept of conflict
> and differences of opinion. There can be circumstances when differing opin-
> ions are equally correct, though they be mutually opposed to each other. It's
> not always necessary for one to be right and the other wrong.   (Trevor, inside
> student)

As Palmer (1993) puts it: ". . . tolerance of ambiguity can be taught as a way
of listening to others without losing one's voice" (p. xviii). Additionally,
understanding the relevance of context is instructive for students in a service-
learning setting because it ". . . help(s) them understand that as human
beings we do have many things in common, yet as a result of how race and
class have situated us within our society, we cannot ignore important identity
differences . . ." (Rhoads 1997, p. 123). Nowhere is this reality more pro-
foundly evident than in prison.

> What a motley crew we made in that little program room at [the prison].
> I often think about the incredible dynamic of our group and wonder what we
> must look like to the people outside that room. People of different colors,
> sexes, ages, education levels, social classes and opinions in a circle, laughing,
> talking, arguing and respecting each other for hours at a time. It has to make
> it difficult for anyone who watches to hold on to the status quo. The status quo
> says that doesn't happen. It says that people are different and that some things
> are never going to change. For two and a half hours every Thursday this semes-
> ter, we proved that untrue.   (Elizabeth, outside student)

A fundamental issue that we discuss at this juncture is that each of us has a culture within which we were raised, comprising our ethnicity, socioeconomic

status, religious beliefs, neighborhood, and many other factors. This culture heavily informs the lens through which we see, experience, and interpret the world. When we are "locked into" a particular cultural perspective—whether that is the culture of the prison, the culture of middle-class America, the culture of the streets, the culture of the "educated"—it becomes difficult to remain open to points of view that are divergent with that perspective. Often, these cultural influences are so deeply ingrained that we are unaware of the depth of their impact on us. By examining our cultural preconceptions, we realize that we overlay these perspectives on everything that we do and say, as well as on how we interact with the world.

> I've been in many settings where I feel poverty, class oppression, racism were all talked about. But somehow it still just felt like words. What is spoken in the . . . class [in prison] strikes me on a much deeper level. (Anisa, outside student)

A further dimension of context refers to the setting in which the service-learning experience takes place, and its effect on everyone involved. In the setting of the prison, for instance, the environment has a tremendous impact on the individuals in the class, the group as a whole, as well as the process of what is transpiring in the course. For example, the simple act of getting into prison each week for class can become inordinately complex and frustrating. The rules seem to have a plasticity that makes it impossible to know what to expect from week to week. What the outside students glean from these experiences, however irksome, puts them in touch with the context—and inherent frustrations—of the setting in which their incarcerated classmates reside. It becomes evident that ". . . the self is inescapably tied to the other. And, perhaps just as important, the social context is the stage upon which the self and the other are framed" (Rhoads 1997, p. 4).

> Walking out of that place every week was hard. It was hard because that was the moment that forced me to face the fact that not all of us were allowed to leave. . . . If prison were anything other that [what] it is, it would be a lot less traumatic to walk out that door and leave someone behind it. (Elizabeth, outside student)

For those imprisoned in the facility, the setting from whence they come and to which they return each class day is authoritarian and oppressive. It is an environment that is antithetical to what is necessary for a productive, creative educational process. Prisons, by their very nature, are restrictive, focused mainly on security and control, rather than the freedom of thought and expression that marks the pursuit of knowledge.

> During our meeting, we're no longer in the jail, we're students of Temple, and no longer . . . are subjected to distractions and restrictions. (Fox, inside student)

The development of this sort of "liberatory" (Shor and Freire 1987, p. 19) environment within the confines of a prison can be viewed as a political act. Further, creating a space in which those who are incarcerated can freely explore issues that directly affect their lives holds the promise of a transformation with far-reaching possibilities. Besides a potential impact on their lives and futures, what both the inside and outside students are able to learn through this exchange is unparalleled. As Shor (1987) points out, "Learning from reality is important, but more than just 'going to reality', you accepted worker-students [in this case, the inside students] as your teachers. That adds a political depth to 'experiential' learning . . ." (p. 30).

> It . . . dawned on me that having a class in a maximum security prison is . . . an act of resistance. By having this class, we are . . . questioning normative assumptions of "criminals" and "criminality" and normative reactions to those who have violated the law. Since society views offenders as immoral individuals who are not entitled to . . . have normal human relations, this class challenges those popular understandings. Every time we to go to [the prison] and have class or even have "normal" interactions with the guys there, we are in fact engaged in an act of resistance. It is a space that humanizes . . . (Diditi, outside student)

One feature of creating a space for individuals to feel free to become engaged involves the arrangement of chairs in the room. This, too, is a political statement in that it reflects how power operates in a group. In a conventional seating structure, power rests in the front of the room with the teacher. In contrast, power is shared when the seats are arranged in a circle. Palmer (1993) describes it this way: ". . . [W]hen the chairs are placed in a circle, creating an open space between us, within which we can connect, . . . we are all being invited to create a community of learning by engaging the ideas and one another in the open space between" (p. 75).

> Class sessions were not "classes" by the usual standard. They were safety nets, zones by which we could come together and discuss issues commonly significant to all of us, problems and solutions that we felt were important to consider and resolve. (Candy, outside student)

Attention and care are critical in fashioning a positive learning environment in which people feel safe. In prison, where a sense of trust is elusive, creating this sort of setting can be a challenge. Additionally, as some of the issues explored are sometimes difficult and sensitive, the group needs to feel that the experience is contained, on the one hand, and unrestricted, on the other. Again, Palmer (1993) sheds light on both of these issues: ". . . [T]he openness of a space is created by the firmness of its boundaries" (p. 72); and ". . . precisely because a learning space can be a painful place, it must have one other characteristic—hospitality. Hospitality means receiving each other, our struggles, our newborn ideas with openness and care" (p. 73).

FINDING VOICE: CHALLENGING THE FORCES THAT IMPRISON

My experience this semester . . . has not only helped me to shed some light on my own prejudices and misconceptions, but it has also strengthened and reinforced my desire to facilitate the process of growing, changing, and realizing the strength and power that is contained in a voice.

Keisha, outside student

When we hold the Inside-Out class with women in the Philadelphia system, it is usually conducted in a large gymnasium, which requires our using a microphone to be heard. During one semester, what occurred as a result became the working metaphor for the class. As participants (both inside and outside students) took the microphone to speak, they backed away from it, uncomfortable with the sound of their voices echoing through the room. We began to explore this phenomenon, and discussed the difficulty, but significance, of finding one's own voice.

In our second class session, . . . we talked about "voice." Our professor, seeing a "teachable moment" in our resistance to the microphone, challenged us to really listen to our voices. She told us that we each have a unique "take" on the world, and said, "No one else has your voice." I think that, as a group, we internalized this message. We listened to ourselves and began to call out each other's voices.   (Catherine, outside student)

Gradually over the succeeding weeks, individuals became more and more comfortable with the sound of their voice, confident that what they had to offer was worth the amplification that the microphone afforded. "This democracy of expression established a mutual atmosphere which encouraged the students to talk openly, not fearing ridicule or punishment for being 'stupid' . . . Very rarely had a professor taken them so seriously, but the truth is that they had never taken themselves so seriously either" (Shor and Freire 1987, p. 23).

At the end of the semester, all participants used their voices with great pride to tell those gathered at the closing ceremony what the semester-long experience had meant to them. This moment was powerful for many who spoke that day, heralding an interior shift that challenged those forces that once had served to imprison and render them voiceless.

PRISON IRONIES: "CORRECTING" A CORRECTIONAL FACILITY

That the conditions found in present experience should be used as sources of problems is a characteristic which differentiates education based upon experience from traditional education. . . . Once more, it is part of the educator's responsibility to see equally to two things: first, that the problem grows out of the conditions of the experience being had in the present, and that it is within the range of the capacity of students; and, secondly, that it is such that it arouses in the learner an active quest for information and for production of new ideas.

Dewey 1997, p. 79

During the final four weeks of the semester, we work on a large project together, from which we produce a report that is submitted to the prison administration, with the hope that some of the ideas will be utilized. In working through this developmental process, we deal with numerous topics, issues, and dilemmas—all grappled with in the context of working on a project with real-world applicability.

For five years, in the women's class, the group designed an ideal correctional facility for women, an idea inspired by plans to build a new institution for women in Philadelphia. [The facility has recently been opened.] Though I am philosophically opposed to further prison construction, it nevertheless served as yet another "teachable moment"—a project through which we examined various issues of concern to incarcerated women within the framework of a real-world example. This exercise provided a challenge to the class to deconstruct the existing model, rethink the philosophical underpinnings of its various dimensions, and reconstruct something totally new and visionary.

During one of the semesters, we realized in our final week that we had developed all of the aspects of the facility except the architectural structure. I asked if anyone had an idea about where we could start, and immediately, one of the inside students, Angela, raised her hand and said she had a design. As I handed over the marker and sat down, she proceeded to sketch a model for the class that, with few changes, became the design adopted by the group. I remember clearly the power of that moment—for everyone involved, certainly myself included. Here was a woman who, after years of incarceration in numerous institutions, extracted from and expanded on her background to lead us exactly where we needed to go. It was a wonderful illustration of the power that can emanate from the crucible of experience.

Five years later, I learned that Angela was enrolled at Temple in art education. Over lunch one day, she told me how significant that moment in class was for her—that it had reawakened her interest in art and given her a new direction for her life. She started taking classes at the local community college, eventually receiving her associate's degree in art. She then transferred to Temple and decided to major in art education, so that she could work with at-risk youth, using art as a vehicle. While continuing her studies, she is now working with young people in creating murals on the walls of buildings in their communities.

TAKING IT FURTHER: ADDRESSING ISSUES OF TRANSITION AND PUBLIC REEDUCATION ABOUT JUSTICE

We have been fortunate. We have made the individual connections and are now faced with the challenge of what to do with our changed views. I don't begin to have the answers. But I have a great deal of hope. . . . I don't know "how". . . but I'm beginning to realize that we must also ask "who"—and I think that together, we have found some of the courage it will take to ask and answer that question.

Catherine, outside student

The Inside-Out experience often is so inspirational that participants want to extend its effect by working on projects that flow from the class discussions. Featured below are two examples of projects that have emerged from specific Inside-Out classes: the first, from a women's class in the Philadelphia system, and the second, from the first class held at Graterford Prison.

In spring 2001, we conducted our third class with women in the Philadelphia Prison System. One of the inside students, Terrie, was paroled during the semester and got in touch with us upon her release. She was interested in continuing the educational process that she had started through Inside-Out. We worked with the relevant people on campus to register her for courses, as well as assisting her in arranging for financial aid. What became clear in the process was the importance of using our resources to support women who are trying to turn their lives around after, in many cases, years of drug use and involvement in the criminal justice system.

Very quickly, the Temple students who took part in the class decided that we had to develop an organized and substantive way to work with women released from jail to address their issues. We met weekly throughout the summer and fall of that year to fashion a program that would accomplish this goal. Over time, others joined in the work, including Temple staff, faculty, and other individuals from outside the university who heard about the project.

The group, which came to be called the Transition Team, offered programming for women in the system focused on goal-setting, decision-making, legal issues, and other topics of interest. We also began to develop a support group on the outside for women who were being released; however, the Transition Team dissolved before that dimension could be fully realized. Looking back now, the completely voluntary nature of this activity made it difficult to maintain. Additionally, as more people joined the group, the focus of the project changed and became somewhat muddled over time. We fell into the trap of wanting to do too much, which led to a sense of burnout and frustration for some in the group.

Nevertheless, there were a number of positive dimensions of the experience. First, the group did some very good work inside the facility for more than a year and a half, and second, the skeletal outline of a program emerged that we now are developing to serve as a follow-up to the Inside-Out class. The program is called "Horizons: The Inside-Out College and Career Exploration Program." This workshop series will serve as a way for inside students to take their rekindled interest in further education and make some concrete plans for themselves around careers and educational pursuits in preparation for life after prison.

> . . . [W]ith enlightenment comes responsibility. We are all responsible. What is the next step after all this dialogue is done?   (Fateen, inside student)

Our first class at Graterford Prison in the summer of 2002 had such a strong connection that the bond created in the class continued beyond the completion of the course. The outside students and the inside students, many of whom are serving sentences of life without parole, initiated an extension program to address the need for public reeducation about crime and justice issues through a series of creative, literary, and media projects. Not merely an academic exercise developed for the course, the project had the potential to be quite long-lasting; indeed, the group, called the "Think Tank," has continued to meet weekly since that time. Hence, theory moved seamlessly into action. Some of the current projects of the group include:

- Theme-based workshops offered to community representatives, public officials, neighborhood organizations, and students, faculty, and staff from area colleges and universities. More than two hundred individuals have attended these workshops to date.
- Production of original Inside-Out writings to be made available through a variety of publications for use in college classes and community groups.
- A series of mural projects in partnership with the Philadelphia Mural Arts Program, SCI-Graterford artists, victims and victims' advocates, and community representatives. The theme of this set of murals is "Victims and Healing."

Most notably, it was this group and the work that we began together that inspired me to apply for a fellowship from the Open Society Institute of the Soros Foundation in fall 2002 to develop Inside-Out into a national model. I received the fellowship and spent September 2003 through August 2004 working in collaboration with a group of individuals on both sides of the wall to create the strategy and materials for replicating the program. Through an initial outreach, we identified 175 instructors from around the country interested in being trained in Inside-Out, and then held our first two National Training Institutes in July 2004 and January 2005 with a total of 29 participants. The Think Tank was involved throughout the entire process, working to refine the core Inside-Out curriculum, developing dimensions of the training program, and serving as an integral part of the training process, by providing both demonstration and feedback sessions to the participants.

Both of these examples are emblematic of what can happen when students experience the inspiration that occurs through a service-learning encounter. Intellectual understanding and analysis of issues combine with concern about and passion for those issues, propelling students to recognize their potential as change agents, ready to take the next step in addressing a particular dilemma. Different from the idea of service-learning as a "feel-good" experience, which can be transient and ephemeral, what we are talking about here involves depth, direction, hard work, and a commitment to make change in the world.

> [This class] has acted as the catalyst in my passion for life and human rights, and was the pivotal point where I realigned my own path. . . . [T]his program

has brought me to a new understanding of life, not just in prison, but in my own life. I have acquired the concrete knowledge of the true interworkings of the system, and at the same time come to realize my own captors in life. (Sarah, outside student)

CHANGE AGENCY: TRANSFORMATIVE EDUCATION REALIZED

I will hold its ideals and values for the rest of my life—but not only to keep them with me but to act consciously with them.

Candy, outside student

According to Palmer (1993), ". . . the central question is whether we are educating students in ways that make them responsive to the claims of community upon their lives" (p. xvii). In providing students the opportunity to interact with the world through guided hands-on experience, and then reflect upon it, we are encouraging them to become agents of change, ready to meet the challenges of an increasingly complex world. In this approach, participants are engaged on many levels, increasing the likelihood of deep, lasting learning—through a blend of theory and practice in a real-world setting—that can have far-reaching implications.

After this course, I realize fully that my theoretical knowledge is only as good as what I do with it. The struggle for me is often frustrating—seeing injustice, trying to change things, sometimes failing and wondering whether anything I can do will make a real difference in people's lives. This course did not eliminate my frustrations. It intensified them. It forced me to look closely at things I may never be able to change. But in facing that, I was able to move past my frustration, to clarify my interests and abilities, and to imagine different ways of being and moving and speaking in this world.   (Catherine, outside student)

Different from charity, service-learning involves a critique of social systems, challenging participants to analyze what they experience, while inspiring them to take action and make change. Transformative events radically shift how we see things. The lens through which we previously had viewed reality is irrevocably altered. It is not just about looking at particular issues from another angle; often, an experience of this kind completely changes the perspective from which one now sees all of life. Thus, service-learning provides both an incubator for and impetus toward social change.

We now carry the torch ! We can provide a voice for those who can't speak, and fight a battle for those unable to fight.   (Abdul, inside student)

After nearly twenty years of going in and out of prison several times a week, and thirteen years of teaching criminal justice, I have become acutely aware of how information about crime and justice is acquired from the media through biased and politicized reports. The complexity of policies that are politically and economically driven cannot be accurately grasped through these sources; in fact, the faulty images of prisons and the stereotypes of the men and women locked inside have a devastating impact on the way our society understands crime and justice. Through our class discussions, we come to see that crime, beyond being a problem in itself, is a symptom of a much deeper social illness—a societal dysfunction in which every one of us, by omission or commission, plays a part. This is what I want my students— those on the outside and those on the inside—to understand: that each of us plays a role in a deep, complex social drama whose unfolding we hold in our hands.

> In our society, we like things simple. We like to turn on the T.V. and pretend that what we see flashing before us is the real world. I know this because that is once how I saw my world. I now realize that this way of life is a danger because it obscures us from the true world in which we intentionally inflict unnecessary suffering onto others. My hope for all people is that they too will one day have the eye-opening experience that I have had in Inside-Out, and if like myself they are inspired to create change, I hope that they do so, because together we can break down the walls, and we can build a better world. (Gene, outside student)

We hold no illusions about the ability of a program of this kind to make institutional or structural changes in either the correctional system specifically or the criminal justice system in general. However, through the transformation that occurs—essentially one person at a time—the seeds of change are sown. Many of those involved in Inside-Out will embark on careers related to crime and justice, while others will work in fields involving the social, political, and economic dimensions of both the public and private sectors. Our hope is that, once in a position to affect policy, these Inside-Out alumni will utilize the new understandings gleaned from this experience to further inform their decisions. It is an approach to social change that takes the long view.

Inside-Out provides an opportunity to put a human face on a problem that can be kept simplified only if it remains faceless. The ability not only to look at issues in complex ways, but also to recognize the complexity in ourselves and others, obviates our propensity for knee-jerk reactions. Stretching beyond our simplistic assumptions will, in time, produce a growing transformation in public thought. By exploring theoretical concepts inside the prison, theory is moved out of the purely mental sphere to a more powerful level—as the mind is engaged, so is the heart. If how we feel, to some extent, drives what we think, herein lies the crux of the transformative potential of this program.

> Every week I get more frustrated. . . . I am ready to act. I tell as many people whose ear I can catch about the prison. . . . I expected to become frustrated and I actually looked forward to it. What I did not fully expect was how heightened my awareness would become. . . . I feel like everyone should be able to see what I see. I guess this increased sensitivity is the best thing that could have ever happened to me. (Sarah S., outside student)

The pedagogy of service-learning has the power to turn things inside-out and upside-down for those engaged in it. It provokes one to think differently about the world, and consider one's relationship to the world in a new way. This approach to learning captures and communicates a dynamism that inspires everyone involved to explore, inquire, and analyze. It is transformative education at its best.

> Through Inside-Out, I have found holes in the walls separating the incarcerated from those that are not. They have been made by individuals inside and individuals outside. It has been a tremendous experience which has begun a process of overcoming ignorance through communication. It has planted a new seed but more still needs to be done. Too few people know or understand what goes on behind the wall. Many people demonize in order to prolong their ignorance, forgetting that above all: everyone is connected by the same fabric of humanity. Comfort can come out of ignorance but Inside-Out shakes things up. It pulls away that comfort zone and creates so many different aspects ranging from disgust to hope. Slowly but surely the walls are coming down. (Diane, outside student)

So, we continue to go behind the walls—to deepen the conversation and transform ways of thinking about crime and justice—from the inside, out. And as we do, the walls grow increasingly permeable. Thus, the isolation and inertia of life behind the walls gives way to transformative power, hopefully turning our lives—as individuals and as a society—inside-out in the process.

> I've emerged from the Inside-Out experience empowered with an unshakable belief in the human capacity to evolve to a higher state of social consciousness. (Paul, inside student and originator of the idea that became the Inside-Out Program)

NOTES

This chapter has been adapted from an article published in the *Michigan Journal of Community Service Learning*, Fall 2002.

I am grateful to a number of individuals who provided editorial recommendations for the original journal article from which this chapter was adapted, particularly Jayne Drake, Mary Hanssens, Emily Miller, Anita Moran, Michael Szekely, and Jill Wolfe. I would like to thank the Philadelphia Prison System and the State Correctional Institution at Graterford for welcoming and supporting this program, as well as the Department of Criminal Justice and the College of Liberal Arts Dean's Office at Temple for giving me the space and encouragement to venture into some new, provocative pedagogical territory. And I am deeply appreciative of the Soros Foundation's belief in this work, expressed through the fellowship awarded me to develop Inside-Out into a national model.

It is of note that Paul, the life-sentenced gentleman whose idea supplied the original impetus for the Inside-Out Program, was transferred at some point after that initial discussion from SCI-Dallas to SCI-Graterford. Early in 2002, when I approached Graterford about the possibility of conducting the course there, he and other men who are doing life were instrumental in making it happen.

Finally, this piece could not have been written without the input of the more than 900 "inside" and "outside" students who have taken part in 19 semesters of The Inside-Out Prison Exchange Program. I have attempted to capture the essence of this experience, and offer these reflections as a tribute to the courage it has taken for them to share their lives with one another and with me. It has been a service to me as a teacher—and a humbling, liberating adventure for me as a person.

REFERENCES

Coles, Robert (1993). *The call of service.* Boston and New York: Houghton Mifflin.

Dewey, John (1997, 1938). *Experience and education.* New York: Touchstone.

Palmer, Parker J. (1993). *To know as we are known: Education as a spiritual journey.* New York: HarperCollins.

Rhoads, Robert A. (1997). *Community service and higher learning: Explorations of the caring self.* New York: SUNY Press.

Shor, Ira & Paulo Freire (1987). *A pedagogy for liberation: Dialogues on transforming education.* Connecticut and London: Bergin and Garvey.

SECTION III

REFRAMING THE INSTITUTIONALIZATION OF SERVICE-LEARNING

12

The Aesthetical Basis for Service-Learning Practice

James Birge

Introduction

The modern college and university is a product of American higher education's historic public purpose. Since the middle of the seventeenth century, American higher education has responded to the needs and demands of society. In 1636, Harvard College was founded in Massachusetts for two purposes. Its first mission was to train clergy to minister to the religious needs of the community. However, Harvard was also founded to educate men who would become public leaders for the emerging commonwealth (Lipset and Riesman 1979; Morison 1937).

By the middle of the eighteenth century, colleges and universities were developing civic leaders who would focus their work on freeing the country from British control. These emerging leaders became the framers of the American Revolution (Snyder 1998).

In the post–Civil War era, the American government created land grant institutions in order to make higher education more relevant to society. The land grant institution accomplished, according to Snyder (1998), "the public work of community problem solving" (p. 10).

During the twentieth century, higher education continued its rich public purpose by contributing to the development of technology, reforming education, and promoting civil rights and antiwar movements (Boyer 1994). Throughout its history, American higher education established itself as a social institution that responded to the changing needs and demands of society and compelled the advancement of the nation. As such, colleges and universities in the twenty-first century are building upon an historic public purpose when they develop civic engagement programs, build campus and community partnerships, and promote service-learning as a teaching method that connects theoretical knowledge with experiential learning and action.

American higher education's challenge is to continue to exercise its rich history of interacting with communities while compelling students to make the connections between the acquisition of knowledge and the broader public good. Without such explicit and implicit connections, we fail to promote democracy. If indeed, a democracy is founded upon citizens deciding what is best for their communities, and if people are disconnecting from the idea of using their knowledge to improve their communities, we are facing an era of continued democratic decline. Although signs of transformation are on the horizon, evidenced by an increase in students' political awareness (Rooney 2003), achieving the outcomes of a civically engaged student population requires the participation of students, faculty, staff, and community partners.

It is encouraging, then, that Campus Compact, a coalition of more than nine hundred academic presidents committed to the civic purpose of higher education, found in its 2003 annual membership survey that, on average, 36 percent of students participate in public service activities—a record high level of engagement. Moreover, the survey revealed that 93 percent of member institutions reported having a partnership with K-12 schools, an increase from 88 percent in 2002. And the mean number of faculty members on Campus Compact member campuses who invoke the service-learning pedagogy increased from 22 in 2002 to 24 in 2003. Likewise, the average number of service-learning courses offered on a campus increased from 30 to 37 during the same time period (Campus Compact 2003).

It is not surprising to learn that the practice of service-learning on college and university campuses has increased. After all, many higher education organizations, disciplinary associations, and federal agencies are providing a second decade of training, technical assistance, and funding to support the practice of service-learning. Research articles and books about service-learning are adding to an expanding culture of evidence that service-learning is a valid and reliable teaching method and form of scholarship. Given this recent increase in the examination of service-learning practice, however, it *is* surprising to see that we have not focused more clearly upon the motivations of faculty who are service-learning practitioners. What is it about service-learning that draws some faculty to be practitioners, but not others? Is there a difference in the practice of service-learning between faculty members who feel that they are *called* to be practitioners and those faculty who invoke its use simply because it is an effective teaching tool?

This chapter *begins* to look at the distinction among faculty practitioners of service-learning who feel that they are *called* to service-learning practice. The chapter puts forth a position that these faculty members view service-learning as a means to integrate not just community service and academic study, but also the broader notions of critical thinking, civic participation, and life-long learning. Throughout my 12 years of working with faculty, I have found that many seek a deeper understanding of service-learning practice, but have found only limited help with their search. I don't offer that statement as a criticism of the work that has been accomplished by me and my colleagues and friends. Indeed, it is because of those accomplishments that we can move toward a deeper understanding of the work. Yet, it is my observation and experience that we now must enhance the training and technical assistance offered to service-learning practitioners with conversations and examinations focused upon what it is that grounds our work personally and professionally. It is my hope that this chapter will either begin or add to conversations about this issue and as a result create a stronger foundation to service-learning practice.

A CHALLENGE TO
SERVICE-LEARNING PRACTICE

Although the findings of the annual Campus Compact survey indicate a steady rise over the last few years in the number of faculty members who integrate community service and academic study, and despite the numbers of professional conferences and association meetings that offer service-learning training and technical assistance sessions, many faculty members continue to eschew service-learning practice. For those faculty who were oriented to the efficiency model of teaching, to provide the most information to as many students as possible within a fixed amount of time, integrating a community component challenges their familiar professional practice. Inserting community-based learning into the formula of traditional teaching practices is, at best, messy. Doing so requires flexibility in teaching practice and course content, creative thinking, and the ability to connect theory to application. Not all faculty members are able or willing to adjust their teaching style to accommodate community-based learning. And not all students are oriented to understanding the social applications of such a disciplined inquiry. As such, service-learning practitioners must recognize and acknowledge that teaching and learning for some of their colleagues and students can only exist within the traditional practice, and that they should always honor those colleagues and students who need or wish to teach and learn through traditional practice. However, for those faculty who see themselves as contributing to leading edge practices in their profession, engaged pedagogical strategies such as service-learning provide a mechanism for integrating their personal call to the profession of teaching with contemporary academic practice, traditional scholarship, and institutional public purpose.

Thoughtful, intentional people center their lives upon sets of values, principles, and practices that compel them to live their lives in particular ways. The manner in which people make decisions about their career, choose their life's partner, find meaning in their lives, decide whether to raise children, or select a place to live in reflect those deep considerations. As intentional people, faculty members arrive at their place in the academy after many years of examining how they will choose to be productive and contribute to society. That process of examination reveals an intellectual curiosity about new ways to teach their disciplines and how to connect the discipline content to contemporary social issues. These faculty members continually seek out engaged teaching and learning strategies for delivering their courses.

In my own work with faculty I have begun to see the emergence of a dynamic that I can best describe as an aesthetical basis for service-learning practice. I am discovering that faculty practitioners of service-learning engage the questions that reveal deeply held reasons for doing what they do, that permeate the logistical approach to professional practice that is embedded in the idea and language of "best practices," and that reveal an inner understanding that integrates personal values with professional activities. To consider a deeper foundation to service-learning practice is to juxtapose the *mechanics* of service-learning practice with the *connection* between service-learning practice and who we are as individuals and how we choose to act in the world. While we need the conversations that lead us to understand how to integrate public service with academic study, we also need to reveal the more deeply rooted understandings that draw faculty toward service-learning practice as an articulation of their belief in the liberating process of education, the role of individuals as contributors to society, and the knowledge that synthesizing distinct learning experiences produces critical thinkers.

Although the "best practices" models may very well imply that the deeper internal conversation is necessary, I do not think that the training and technical assistance available to service-learning practitioners offers as much support to that conversation as it does to the element of syllabus design, for example. This nascent idea of an aesthetical basis for service-learning suggests that the faculty member teaches with a specific purpose in mind that transcends the sharing of knowledge and reveals a comprehensive understanding of the intersection of knowledge, experience, and self-understanding. This concept of the purpose of teaching and learning rests on the idea that the authority of personal experience compels the student to learn more deeply and to apply knowledge more broadly. When this synthesis occurs we realize that the student and faculty member sustains a personal and professional ethos of engagement with the discipline, with learning, with communities, and with oneself. Such a relationship between faculty member, student, knowledge, application, and self reveals a more fluid existence of faculty members as teachers *and* learners, students as learners *and* teachers. Through such an approach, faculty members invoke a more authentic concept of education that does not presuppose a rigid understanding of content and impose that upon students. Rather, this reciprocal learning dynamic allows students and community partners to take knowledge presented by the faculty member, synthesize it with other knowledge, and apply it to work and life. Palmer (1983) suggests that this kind of interaction with knowledge—the awareness of how to integrate knowledge with oneself, life, and work—reveals a deeper understanding of our own identity. "At its deepest reaches," he writes, "education gave me an identity as a knower. It answered the question, 'Who am I?' by saying, 'You are one who knows.' The knowledge I gained through education was more than a tool for my vocation; it became a source of my self-understanding as one whose nature it is to know" (pp. 20–21).

What Palmer reveals in his statement is the power of integrating knowledge with the understanding of our own existence. Knowledge in this instance is something more than a career-enhancing device, more than a phenomenon to be admired, more than collections of data and information to be applied.

In its most authentic role, knowledge forms the basis of who we are as individuals and how we choose to live our lives in a complex world. An understanding of this role of knowledge underpins the practice of service-learning as an extension of a faculty member's personal commitment to teaching, while reflecting a deeply rooted motivation for connecting knowledge, experience, and self-understanding. Service-learning practitioners who engage a process of integrating knowledge with experience discover that such practice yields more than having students develop a deeper understanding of course content and application of theory to practice. Combining academic and experiential learning yields a deeper understanding of oneself and the world.

In this understanding of service-learning practice, as an extension of an internal "conversation" about one's own existence, the faculty member acknowledges a process of making the connection between his or her own encounters with the world and how those encounters shape future action. Thus, the basis for future action is grounded in the individual's prior encounters reflecting the individual's fundamental identity. This process of encounter and realization reveals the world anew, making explicit the connection between identity and action. Palmer (1969), in his treatment of the theory of understanding, reveals that the phenomenon of aesthetics integrates the world of encounter with the world of self-understanding such that "we see the world 'in a new light'—as if for the very first time" (p. 168). Therefore, the aesthetical basis for service-learning draws us deeper into the fundamental reflections of how to act in the world, how to integrate self-understanding, experience, and knowledge, and how to connect identity with work.

It is important here not to conflate the idea of reflection as an instrument for connecting experiential and academic learning with the idea of a "fundamental reflection" about what we teach as a reflection of who we are. The former is a device such as journals, guided discussions, and purposeful reading assignments that creates a link between the course objectives, disciplinary theory, and class discussions with community-based encounters. The latter concept of reflection is a revelatory experience that presupposes in our teaching a deeply held commitment to bring one's own skill and passion to bear on improving the lives of people in society. This self-analysis of who we are, why we teach, and how we link the two holds the power to reveal the phenomenon of integrating personal ideals and values with professional work. Yet, few opportunities exist for us to begin to examine this phenomenon. I suspect, however, that many of us are looking for a way to begin this examination, for a way to respond to the question that many service-learning practitioners and critics ask—why do you integrate community-based learning with classroom-based academic learning?

A few years ago I was meeting with a group of faculty from a local university when one of the faculty members said to me, "I love service-learning, it is so easy to do." The statement dumbfounded me. I never found service-learning to be easy, and most of the faculty I have spoken with about service-learning practice frequently comment on the difficulty of integrating public service and academic study. My initial thought about this faculty member's statement was that, since I found service-learning practice to be difficult, I must have been doing something wrong. I frequently recall this faculty member's remark about service-learning. On one hand, it may be the case that this faculty member has perfected the technical elements of service-learning practice—preparing students for community-based work, developing partnerships with community-based agencies where students will perform appropriate work, and designing reflection activities that connect experiential and academic learning—and, therefore, does not need to spend time and energy creating partnerships, redesigning syllabi, or selecting course texts that complement the integration of theory and practice.

It may also be the case that this faculty member has discovered and reconciled what makes service-learning practice so difficult and what still eludes so many of us. It is a discovery of what I think many of us seek, but do not know how to proceed—a discovery that, once realized, exposes our own vulnerabilities and projects them for others to see—others who evaluate our work and ultimately, in the case of service-learning practitioners, evaluate our identity. The discovery is that service-learning practice reveals our dis-ease with the way the world is and acknowledges that our humble and noble attempts to bring about change will be implemented by others of another generation. Yet this realization that our efforts to create change will be carried out by others does not deter the reality that it is, nonetheless, thoughtful intellect that creates change. The effort to connect a sense of identity with improving the world through work is not a futile attempt to place one's mark in the world. Rather, it is an historic exercise in creating meaningful change, albeit countercultural in contemporary society. Here, I am reminded of the frequently cited quote attributed to the anthropologist Margaret Mead, "Never doubt that a small group of thoughtful, committed citizens can change the world; indeed, it's the only thing that ever has."

What makes service-learning so messy is that we are creatures of curiosity always seeking a place that we know is an elusive and almost impossible place to find—peace. In his treatise on spirituality, Rolheiser (1999) states it more eloquently than I can:

> Put more simply, there is within us a fundamental dis-ease, an unquenchable fire that renders us incapable, in this life, of ever coming to full peace. This desire lies at the center of our lives, in the marrow of our bones, and in the deep recesses of the soul. We are not easeful human beings who occasionally get restless, serene persons who once in a while are obsessed by desire. The reverse is true. We are driven persons, forever obsessed, congenitally dis-eased, living lives, as Thoreau once suggested, of quiet desperation, only occasionally experiencing peace. (p. 3)

Yet we move forward, always forward, seeking to find meaning in that inexorable pull toward integrating theory, practice, and articulating our identity through our work. It is a legitimate and worthy process offering the potential to reveal the underpinning foundation to our own identity and what it means to share that identity and purpose with others through our teaching.

However, herein lies the paradox of the aesthetical basis for service-learning. Whereas the primary responsibility of faculty is to make the unknown known, few, if any opportunities exist for faculty practitioners of service-learning to make known to themselves their own understanding and motivations for integrating service and academic study. Although there do exist isolated populations of faculty practitioners who have engaged the thorny questions of their own existence and motivations for teaching, they have not done so as a result of the literature, training/technical assistance, or conferences on service-learning. Moreover, at many institutions, dominant academic culture challenges the idea of an aesthetical basis to service-learning practice. This is not to say that institutions are explicitly hostile to the idea of a more intentional practice of teaching. On the contrary, I suspect that all colleges and universities would welcome faculty who have excelled in their field and who have cultivated a deep relationship between the development and sharing of knowledge in order to improve the lives of others. Yet, all too often institutions do not make space for faculty members to examine their own teaching and what it means to them, their students, and their communities. Between heavy teaching loads, publication expectations, student advising responsibilities, and committee meetings, faculty members are hard-pressed to find time and space that allow for a meaningful examination of integrating self, teaching, and knowledge. And, as hooks (1994) notes, "Many professors remain unwilling to be involved with any pedagogical practices that emphasize mutual participation between teacher and student *because more time and effort are required to do this work*" (p. 204, emphasis added).

It is too simplistic to suggest that the way to develop opportunities for faculty to examine their teaching is to merely encourage faculty to gather around a lunch table and start talking. The field of service-learning practitioners needs and deserves more structured opportunities to engage the deeper questions of how their professional practice is a reflection of their self-identity. Although opportunities exist for faculty to be trained on service-learning practice, rarely do those workshops and symposia contain opportunities for deeper reflection on the important issue of self-identity and how our practice of service-learning evinces the notion of commitment to the struggle to find and promote a better world. However, I must acknowledge that to provide such opportunities is no easy task. Tight budgets, federal mandates, limited free time, and the incessant drive to quantify impacts of service-learning (how many participants were involved, how many hours were served, how many faculty teach service-learning courses, etc.) challenge the ideal of providing the length of time, space, and dialogue that compels a free-flowing exchange of ideas and thoughts on our self-understanding and identity with professional practice. As constraining factors to such dialogue recede, we need to take advantage of the opportunities to create the examinations of our work so that we provide a more compelling educational experience for our students and a life-giving environment in which we work.

I am confident that the day is on the horizon when we will see multiple, structured opportunities to bring our dis-ease to a level of understanding and appreciation and that compels more grounded professional practice. Until that day arrives however, we do not have to struggle alone with our deeper questions. My experiences talking with faculty suggest that there are many of us who are in the process of seeking the deeper meaning of, and why we are called to, service-learning practice and that all of us are at different levels in this process. Some are just starting to ask the question, "Why is it that, despite the difficulty with integrating academic study and service, I continue to do it?" Others have found a path toward their own truth and ask the question, "If this is who I am, how do I either affirm or deny myself in my professional practice?" While still others have found their way toward integrating self and professional practice and ask, "How do I expand upon this understanding and help others find their own truth?"

The lack of networks of people who examine this idea of the aesthetical basis to service-learning or the lack of workshops on this topic at professional conferences can not deny that there are indeed service-learning practitioners who are asking themselves these fundamental questions. In order to build a broader, stronger base to service-learning practice we must be willing, as service-learning practitioners, to move beyond our isolated thinking about our call to professional practice and move toward open conversations with colleagues about our questions. Such a move affirms our struggle to legitimize service-learning within the academic culture and confirms that service-learning is an outcome of self, theory, knowledge, and practice.

Without this deeply seated concept of why it is that we integrate academic and experiential learning, we risk making experience tangential to the academic objectives of the course and, therefore, disconnecting knowledge and experience. Although keeping the experience disconnected from the course creates more space for course content, it also reduces the opportunities for shared knowledge and collaborative learning. Huberman (1993) highlights this risk when he writes, "One finds, of course, the same phenomenon among superior artists, athletes, craftspeople, mechanics, and even surgeons: levels of automaticity that serve the important purpose of freeing cognitive space for more complex operations but that can also reduce articulate awareness and explication of one's behavior to mush" (p. 17).

CONCLUSION

Much of the expansion of service-learning practice is due to the multiplicity of conferences, workshops, training sessions, publications, and consultants that focus on the pragmatic elements of integrating community service and academic study. These "pragmatic elements" include such things as syllabus design, reflection activities, assessment devices, partnership development strategies, etc. All are critically important elements for the *practice* of service-learning. And, indeed, we need to continue to provide these enabling mechanisms so that service-learning practice continues to expand, deepen, and broaden at academic institutions and in communities. Yet, without addressing the aesthetical underpinnings to our practice of service-learning, we may be building the structure of service-learning that lacks a deeper connection to the fundamental reasons for the work, and ultimately disables the foundation and sustainability of service-learning.

Service-learning offers the promise of allowing higher education institutions to articulate their missions, to engage students more deeply in the learning process, to develop meaningful relationships with their host communities, and to educate men and women to take leadership roles in a changing world. It also offers the opportunity for faculty to find deeper meaning in who they are, why they teach, and how to bring their personal and professional insight to bear on society. To realize the promise and opportunity that service-learning offers demands much work and energy, however. Administrators, faculty, staff, community partners, and students must commit their leadership, innovation, courage, and persistence to the work in order for service-learning to have firm purchase on the higher education landscape. To have a meaningful, rigorous service-learning program requires people to move outside of that with which they are most familiar and step into the unknown. It is a courageous act. However, just as "gardens are not made by sitting in the shade," according to Rudyard Kipling, neither will we make change by practicing our work solely in the shadow of the ivory tower.

REFERENCES

Boyer, E. L. (March 9, 1994). Creating the new American college. *The Chronicle of Higher Education*: A48.

Campus Compact (2003). 2003 Service statistics: Highlights of Campus Compacts annual membership survey. Providence, RI.

hooks, b. (1994). *Teaching to transgress: Education as the practice of freedom.* New York: Routledge.

Huberman, M. (1993). The model of the independent artisan in teachers' professional relations. In J. Warren Little & M. Wallin McLaughlin (Eds.), *Teachers' work: Individuals, colleagues, and contexts* (pp. 11–50). New York: Teachers College Press.

Lipset, S. M. & Riesman, D. (1979). *Education and politics at Harvard.* New York: McGraw-Hill.

Morison, S. E. (1937). *Three centuries at Harvard.* Cambridge, MA: Harvard University Press.

Palmer, P. (1983). *To know as we are known: A spirituality of education.* San Francisco: Harper Row.

Palmer, R. (1969). *Hermeneutics.* Evanston, IL: Northwestern University Press.

Rolheiser, R. (1999). *The holy longing: The search for a Christian spirituality.* New York: Doubleday.

Rooney, M. (January 31, 2003). Freshmen show rising political awareness and changing social views. *The Chronicle of Higher Education*: A35–38.

Snyder, R. C. (1998). The public and its colleges: Reflections on the history of American higher education. *Higher Education Exchange*: 6–15.

13

Putting Down Roots in the Groves of Academe: The Challenges of Institutionalizing Service-Learning

Matthew Hartley, Ira Harkavy, and Lee Benson

Introduction

Service-learning, by almost any measure, has been an enormously successful academic innovation (Stanton, Giles, and Cruz 1999). A mere two decades ago Campus Compact was founded by three university presidents intent on making service an integral part of their students' experiences and aspiring someday to have 100 college and university presidents as members. Today, more than 900 presidents and their institutions have joined and 30 state offices provide training and technical assistance to students, faculty, and administrators in the areas of service-learning and civic engagement.[1]

However, service-learning has been embraced by institutions to varying degrees. On some campuses, proponents are located in "service enclaves" (Singleton, Burack, and Hirsch 1997)—isolated faculty members laboring with meager support. On other campuses, service-learning courses are found in only a few disciplinary areas—professional programs, the social sciences and a few environmental courses from the life sciences thrown in for good measure. However, some campuses have managed to embed service-learning into institutional life to a striking degree (Holland 1997). At these institutions, an influential group of faculty members, administrators, and board members have embraced the pedagogy and worked collectively to engage members of the larger institution in discussions about institutional priorities and purpose. Institutional support is evident in their annual operating budgets, the allocation of staff lines, and in some cases by the creation of a central office supporting these initiatives. A few colleges and universities have amended admissions requirements or redefined faculty roles and rewards and community-based research has gained some currency as a legitimate form of scholarship. But why the disparate levels of commitment? What are the factors that influence degrees of institutionalization and what are some of the strategies institutions have employed to promote service-learning?

The purpose of this chapter is to examine these questions by means of a review of the literature on institutionalization and by sharing the experiences of individuals at four campuses, Swarthmore College (Swarthmore, PA), Tufts University (Medford, MA), the University of Pennsylvania (Penn) (Philadelphia, PA), and Widener University (Chester, PA). The research and analysis reported here represent the pilot phase of a larger research project examining how colleges and universities embed civic engagement efforts into the lives of their institutions. What little quantitative data exists (e.g., Campus Compact's 1998 survey, discussed below) suggests that many—perhaps most—colleges and universities have enjoyed only partial success institutionalizing such initiatives. Although a few researchers have done important preliminary work identifying factors that promote or impede institutionalization (Holland 1997; Ostrander 2004; Ward 1996), we contend that for service-learning to firmly take root in the academy, advocates must learn to identify both the structural and ideological (or cultural) features of their own institutions if they wish to devise effective strategies for addressing them.

These four private institutions were chosen in part because they represent distinct institutional types: Swarthmore is a liberal arts college, Widener a regional comprehensive institution, Tufts a small research university and Penn a large research university. Although some students and faculty members have been involved in community-based learning for a decade or more at all of these institutions, comprehensive structural support for these efforts is more long-standing at Penn and Tufts and quite recent at Swarthmore and Widener. In short, these institutions represent different points on an institutionalization continuum.

Consistent with the practices of good research (Glesne and Peshkin 1992; Maxwell 1996; Rossman and Rallis 1997), it is important to note the relationship of the three authors of this study to Penn's Center for Community Partnerships (CCP) and to justify its selection as a site. Ira Harkavy directs CCP and has for years collaborated with Lee Benson in the classroom and on numerous articles and chapters. Benson and Hartley both serve on CCP's faculty advisory board. However, in order to mitigate bias, the decision was made to have Hartley, a second-year assistant professor whose relationship with CCP is quite recent and who knows none of the other participants (with the exception of Harkavy who was interviewed as the director of CCP), conduct the interviews. Two people were interviewed from each campus, the person responsible for day-to-day operations of the service-learning or community engagement initiative (e.g., director, dean, vice president) and a senior administrator or a senior faculty member. Their accounts were supplemented (and confirmed) through an analysis of institutional documents (e.g., reports, brochures, syllabi, Web pages.) Each interview was audiotaped and subsequently reviewed. Extensive interview notes were drafted and quotes that captured particularly important points were transcribed verbatim. The entire set of interview notes was coded, which resulted in the development of the key themes presented in section III.

We included CCP in the study for several reasons. First, it is one of the relatively few long-standing programs of its kind and two of the authors (Harkavy and Benson) have an intimate familiarity with the program and Penn. As such, it is an information rich source (Patton 1990) that seemed a shame not to use. Second, we had a professional interest in comparing and contrasting CCP's experience with that of other institutions. Third, the data drawn from this limited sample was intended primarily to illustrate the empirical and theoretical work found in the literature. Thus, the concerns we had about potential bias were mitigated by the fact that this is a pilot study and its findings are therefore tentative and exploratory. However, we now hold that the themes that emerged yield insights that not only are worthy of future exploration but that, because of their consistency across the interviews, are trustworthy and may prove useful to practitioners.

DESCRIPTION OF THE Four INSTITUTIONS

Founded in 1992, Penn's CCP is a long-standing experiment in community-based learning and civic engagement.[2] CCP's director, Ira Harkavy, is associate vice president and reports to the vice president for Government, Community and Public Affairs with an indirect reporting relationship (a dotted line on the organizational chart) to the Provost. Harkavy manages a staff of 15. The Center has three advisory boards—community, faculty, student, and an external board that helps the Center's staff establish priorities and secure external support. Although Penn provides space and overhead, the university's budgetary strategy of "every tub on its own bottom" (Resource Center Management) means that the center is dependent upon grant money and fundraising to support many of its activities.

CCP helps to facilitate three kinds of activities: academically based community service, direct traditional service, and community development. Its projects include partnerships with West Philadelphia public schools, training programs for local school teachers around service-learning, a community arts initiative, among others. CCP also supports the many faculty members engaged in service-learning across the university. In 1992, 11 service-learning courses were offered. Today, over one hundred and twenty courses from a wide range of disciplines link Penn students to the community.

Tufts University's institutional commitment to civic engagement is underscored by the creation of the University College of Citizenship and Public Service in 1999 by the board of trustees thanks to the efforts of former president John DiBiaggio. University College provides a formal structure for coordinating the civic engagement efforts that have been occurring at Tufts for many years. Under the leadership of its dean, Robert Hollister, the stated goal of University College is to make "active citizenship a hallmark of a Tufts University education." University College sponsors a range of activities including voter registration drives and lectures and symposia on political issues and the presidential candidates. The Public Service Scholars program selects approximately fifty students each year whose role is to serve as catalysts for promoting service activities among students and faculty. The Faculty Fellows program awards two-year grants to a select group of tenured faculty members. "Fellows are selected because of their potential to engage other faculty and infuse Education for Active Citizenship throughout the University."[3]

Swarthmore College offers something of a contrast. Although senior administrators point to a long-standing ethos of service, community engagement efforts received little or no formal institutional support until relatively recently. In the late 1980s, a "Green" Dean—a recent Swarthmore alumnus/a—was hired to help coordinate student service efforts. However, faculty engaged in service-learning had to go it alone.

A series of events in the 1990s altered matters: the hiring of President Al Bloom in 1991 (whose scholarly work centers on "ethical intelligence"); an initially ill-fated effort to create a multicultural curriculum, which sparked a series of debates about Swarthmore's educational mission; and a new capital campaign all fed an emergent interest in civic engagement and service. In 2002, the Lang Center for Civic and Social Responsibility (conceived of and funded by former board chair Eugene Lang) was founded to provide a broad umbrella under which multiple efforts could be coordinated and supported.

Finally, Widener University is an institution poised on the brink of change. Its new president, James Harris, was hired by the board after his successful tenure at Defiance College where he led an institutional transformation revolving around civic engagement. Widener is a regional comprehensive institution with traditional disciplinary offerings as well as a number of professional programs. A series of presidential initiatives have recently been launched. A new Office for Community Engagement has also been instituted and its director reports directly to the president. Harris is also leading an effort to develop a closer working relationship with the wider Chester community. A small group of faculty had already been involved in service-learning and Harris is hoping to augment that group.

INSTITUTIONALIZING SERVICE-LEARNING: LESSONS FROM THE LITERATURE

Service-learning has found its way into many, perhaps a majority of American college and university campuses. Data from a 1998 Campus Compact survey of 300 member campuses found that "99% of the respondents reported having at least one service-learning course, up from 66% in 1993. Of the 99%, 19% had 40 or more courses, 48% had between 10 and 39 courses, and just 33% had less than 10 courses" (http:www.compact.org/faculty/specialreport. html). Between 1998 and 2002, the overall percentage of faculty undertaking service-learning on member campuses grew from 13 percent to 22 percent (Hartley and Hollander 2005). Campus Compact describes this state of affairs with a three-tiered "service-learning pyramid." The vast majority of institutions, which form the foundation of the pyramid, have a small percentage of faculty (between 0 and 10 percent) engaged in service-learning, few institutional support mechanisms and the net result is sporadic and uncoordinated community engagement. A second (and smaller) group of institutions report that 10 to 24 percent of their faculty members are using service-learning in at least one class. These institutions tend to have offices dedicated to supporting service-learning or civic engagement (although Compact's report does not attempt to measure the adequacy of these structures by, say, comparing numbers of staff with the size of the faculty). Respondents report that the chief academic officers (e.g., provosts, vice presidents for academic affairs) at these institutions support service-learning through various incentives, encourage the development of new courses by offering stipends or a course release, but faculty roles and rewards remain unchanged. Fully engaged colleges and universities represent the tiny apex of the pyramid—institutions where service-learning is valued across all disciplines and specific policies have been adapted (e.g., faculty hiring and promotion and tenure guidelines) to support the efforts. In sum, though service-learning as a pedagogy is widespread, the extent to which it has been woven into institutional life varies considerably among Campus Compact member institutions.

Attributes of Higher Education that Impede Service-Learning

There are a number of factors mitigating these efforts. First, colleges and universities are "loosely coupled" organizations (Weick 1976). A college or university is divided into various schools, which are divided further into divisions and then again into departments. This academic specialization is the legacy of the German research university model and though it has historically proven to be a useful strategy for developing new bodies of knowledge, it has splintered the faculty. Further, power is diffused in loosely coupled systems. Because the units have specialized knowledge (the board isn't about to tell the economics department what to teach) change must occur through discussion and persuasion rather than command. Though the board of trustees has broad powers—the president serves at its pleasure and the board has the authority to create or close entire academic programs—it relies on the cooperation of administrators who understand the institution's inner workings and the faculty, whose expertise must be brought to bear if there is to be any curricular change.

A second factor impeding change is that people are busy with other things. Change requires additional time and effort and on any campus, people's attention is the scarcest resource (Hirschhorn and May 2000). This is particularly true of curricular initiatives. Faculty members teach, research, advise, serve on standing committees, write letters of recommendation, mentor young scholars, participate in peer review for academic journals and much more. It is therefore no surprise that faculty members see themselves as having precious little time to pursue any activity whose purpose may be construed as tangential to these core duties.

This brings us to the third mitigating factor, Platonization. Plato advocated the search for theoretical knowledge as the primary end of the academy and the dead hand of Plato is felt even today (Benson, Harkavy, and Hartley 2005). Academic disciplines tend to devalue research whose purpose is addressing local problems. The system of peer review tends to reward scholarship in familiar forms. Of course, Ernest Boyer's (1990) idea of a "scholarship of application"—using disciplinary knowledge to address particular community concerns—has sparked discussions at many institutions, but it has gained currency in comparatively few, particularly research universities. Given this rather limited definition of scholarship, small wonder that faculty members working toward promotion and tenure are reticent about becoming involved.

Finally, some scholars question the very propriety of promoting civic engagement. They argue that the primary purpose of higher education is to encourage the development of analytical skills, facility in written and oral communication, and knowledge of a particular field of inquiry. What students choose to do with this knowledge (or whether they do anything at all) is, from such a perspective, immaterial and beyond the scope of higher education. The idea of value neutrality is a potent inhibiting force. The German university model and its ethos of "value freedom" in research heavily influenced academic norms and helped to de-emphasize higher education's role in shaping students' values.[4]

Historians Lee Benson and Ira Harkavy (2002) note that

[although] "value-free" advocates did not completely dominate American universities during the 1914–1989 period . . . they were numerous enough to strongly reinforce traditional academic opposition to real-world problem-solving activity, and they significantly helped bring about the rapid civic disengagement of American universities.   (p. 13)

What Promotes or Impedes the Institutionalization of Service-Learning?

Several scholars have attempted to isolate the factors that promote or impede the institutionalization of service-learning. Kelly Ward (1996) examined five institutions whose institutional rhetoric asserted their commitment to service. She concludes that substantive commitment is indicated by: (1) the establishment and funding of a service-learning office; (2) broad-based discussions by faculty members about how to appropriately incorporate service into the curriculum; and (3) the tangible and symbolic support of institutional leaders. Barbara Holland's (1997) analysis of 23 institutional case studies supports and extends Ward's findings. Holland identifies seven distinct factors that indicate a commitment to service: (1) an institution's historic and currently stated mission; (2) promotion, tenure, hiring guidelines; (3) organizational structures (e.g., a campus unit dedicated to supporting service activities); (4) student involvement; (5) faculty involvement; (6) community involvement; and (7) campus publications. Holland goes one step further, however, and for each factor she differentiates between the activities of institutions for which commitment to service is of low relevance, medium relevance, high relevance, or those for whom service is fully integrated. The resulting matrix paints in broad brushstrokes a picture of what institutionalization entails. Holland underscores that the matrix is descriptive not prescriptive. "Without further research, the relationship, if any, among the levels of commitment to service is not clear, especially when one considers that movement could be in any direction on the matrix" (p. 40).

Although identifying such factors is an invaluable starting point for discussing the challenges and opportunities shared by service-learning proponents, Susan Ostrander's (2004) investigation of civic engagement efforts on five campuses underscores how particularistic the challenges can be from campus to campus. Ostrander argues four essential points. First, the emphasis of any given civic engagement initiative differs from institution to institution. One campus may see its central project as making curricular change while another may emphasize the importance of building equitable partnerships with community organizations. Second, certain "local factors" support or impede civic engagement—a distinctive mission, a consensus that the curriculum needs to be revitalized, a faculty that is actively involved in institutional decision-making—such unique elements must be identified and addressed for institutionalization to succeed. Third, a successful initiative requires the development of a convincing intellectual rationale that explains how the effort will result in better teaching and scholarship. Fourth, new organizational structures must be built to sustain the initiative and these structures not only must support the work within the campus (i.e., the faculty and students) but also serve community partners.

The findings of each of these researchers echo the earlier work of organizational theorists Paul S. Goodman and James W. Dean (1982). Goodman and Dean offer a framework for describing how any new organizational behavior comes to be institutionalized. They delineate five stages: The process begins when people become aware of a new activity or behavior—someone tells them about it and explains its value. In the second stage, a small group of individuals try the new behavior. Their experimentation yields important information about how valuable and viable it is in that specific organizational context (i.e., Does it work and do others find it acceptable or tolerable?). If the new behavior turns out to be more satisfying, effective, or enjoyable than its alternative (or if it attracts positive attention from valued peers or superiors), more people will try it and some individuals will begin *preferring* the behavior. If enough individuals, either a majority of people within the organization or the majority of influential people who control roles and rewards, come to prefer the behavior, then a new institutional norm is established. A consensus emerges that the behavior is appropriate and valuable. Institutionalization is complete when people within the organization view the behavior as an expression of the core purpose of an institution. "This is who we are."

Notice that what Goodman and Dean (and Ward, Holland, and Ostrander) are all alluding to is that institutionalization is the product of both structural and ideological change. Structural elements (more resources and more policies) are alone insufficient to alter the day-to-day behaviors of individuals, particularly those working in loosely coupled organizations like colleges and universities. Conversely, no band of zealous advocates for any idea will spur broad-based change if they cannot secure adequate resources. Structure and ideology are the twin drivers of institutionalized change.

The expansion of service-learning and the burgeoning emphasis on civic engagement at many colleges and universities has been likened to a social movement (Hollander and Hartley 2000; Maurasse 2001). What we argue here is that although social movement theorists have tended to stipulate that movements do not occur in bounded organizational contexts (Diani 1992), in fact, there appear to be striking similarities between social movements and organizational change at colleges and universities where, because they are loosely coupled organizations, hierarchical and bureaucratic constraints are of limited influence. Like a social movement, the idea of service-learning tends to be embraced by a small group of proponents. Some are "true believers"—individuals who are absolutely convinced of the efficacy (and even righteousness!) of their cause (Hoffer 1951). Others embrace the idea for pragmatic reasons (i.e., the new idea for a better way of doing things). These individuals form the ideological base of the movement—they are the ones who initially formulate the rationale for the innovation and energetically advocate for it. If influential organizational members form part of the group (e.g., members of the board, the president and administrators, senior faculty), this "guiding coalition" (Kotter 1996) also serves as a powerful political base. The success of the group is dependent on many things, but, broadly put, they must find a way to gather resources (as expressed in Resource Mobilization Theory, perhaps the dominant framework in social movement theory) but they must also secure the support of others—converting them to the cause or at a minimum convincing them not to oppose it.

INSTITUTIONALIZATION OF SERVICE-LEARNING: LESSONS FROM THE FIELD

In this section, we compare and contrast the factors that have influenced the institutionalization of service-learning at four institutions. The subsections below to a degree correspond to Goodman and Dean's framework in that we are essentially describing how ideas spread, how behaviors proliferate and come to be preferred and eventually how an innovation achieves normative status. We make no claim that these findings (based on rather limited data collection at four private institutions) are generalizable to all institutions of higher learning. Our purpose is simply to underscore that structural and ideological elements both are required to produce deep institutional commitment and to examine some of the disparate challenges in achieving them.

Pre-change

The seeds of change can long lie dormant in an organization before innovation occurs (Kanter 1983). A concerted community engagement effort did not come to full flower for many years at these four institutions for a variety of reasons. At Widener, the concept of service to the community had been somewhat constrained. The institution sent its corps of cadets to march at city celebrations but over time, Chester's daunting crime and poverty resulted in the institution isolating itself. For example, the university routinely told its students never to cross the highway into the nearby neighborhoods. Penn's relationship to West Philadelphia several decades ago was strikingly similar. One trustee apparently suggested building a wall around campus to keep students safe.

Conversely, the beauty of Swarthmore's surroundings somewhat mitigated the urgency to serve, though Chester is only a few miles away from that campus. Further, as one participant noted: "Some people had gotten burned in Chester in the past." Creating service projects proved challenging and in a few instances there were uncomfortable interactions with members of the community who questioned the motives of faculty members from an elite, wealthy, private college.

Despite these challenges, there were a few early innovators and pioneers engaged in the community—faculty members who had found ways to incorporate this work into their teaching and scholarship and student volunteerism. For example, at Swarthmore several faculty members had successfully developed partnerships with community based organizations and schools. Also, student organizations had for decades organized service projects, some of them quite ambitious. Some twenty odd faculty members at Widener involved in service-learning began meeting to share ideas and promote the practice fully a year before President Harris was hired. The experience of these people provided a latent fund of knowledge about local community needs and the logistical and pedagogical imperatives of the work. Such individuals may also represent a potential coalition of proponents committed to such work. Certain values already have a degree of resonance within a community. A commitment to social justice stemming from its Quaker roots forms an essential part of Swarthmore's mission, for instance.

Introducing the Idea

Until an innovation is fully institutionalized (i.e., until the behavior is found throughout the organization, is supported by adequate structures, and is

widely viewed as expressing core institutional values), a continual effort must
be made by proponents to introduce the idea to others. Various venues of
communication were used at these institutions—presidential speeches, town
meetings, workshops, seminars, and committee meetings. Although it is certainly true that anyone has the potential to instruct or inform another person, his or her influence in the organization varies considerably. For example,
a faculty member may draw a departmental colleague or students into a conversation about community engagement. A chair, because she controls the
departmental agenda, may be able to facilitate a conversation among a larger
group of faculty and a dean even more so. Senior administrators (both academic and administrative) have access to most if not all of the influential people on a campus. However, the person able to command the greatest number
of people's attention is clearly the president (Schein 1985).

President Harris arrived from Defiance College intent on realigning the
institutional mission to serve the community. One administrator remarked:
"President Harris shared with me that when he was interviewing for the
position he told the board 'don't hire me unless you are committed to
transforming this institution around the idea of civic engagement.' " Once
hired, Harris held a series of town meetings to discuss this vision and, more
recently, has led an effort to redraft Widener's mission statement. The first
sentence now states: "As a leading metropolitan university, we achieve our
mission at Widener by creating a learning environment where curricula are
connected to societal issues through civic engagement." Although presidents
can lead conversations, they do not control them. At Swarthmore the
emphasis on civic education came about in a more roundabout way.
President Al Bloom arrived on Swarthmore's pristine suburban campus with
an interest in promoting "ethical intelligence" in an increasingly global
world. With his encouragement as well as the provost's, a group of faculty
crafted a new multicultural curriculum that encountered fierce resistance.
This setback, however, led the provost to invite faculty members to form
seminar groups to discuss multiculturalism and the curriculum. These
ultimately led to a range of discussions about the college's mission and a reaf-
firming of values constant since its founding by Quakers, a commitment to
social justice and service to the community. Without the firm commitment
of the senior leadership, including the president, engaging in a broad-based
discussion of core issues such as curricular reform is difficult indeed.

Such conversations serve a number of purposes. At their most fundamental,
as Goodman and Dean might note, they are a means of introducing ideas.
Through collective debate and discussion, they allow ideas to be vetted and
enable some to gain broad currency. Important distinctions need to be
drawn: What do we mean by "service," who is our community, and what
are the characteristics of civic or democratic behavior? Are we hoping to
promote public service, participation in service-learning, community-based
learning, or real-world problem solving (our preference)? At times these
conversations were difficult and uncomfortable. One administrator at
Swarthmore explained: "It's like having the sand inside the oyster—there's

going to be a lot of itching and scratching, lots of reflection." Proponents of service-learning had to respond to a range of objections at each of these institutions. Faculty members unfamiliar with the pedagogy questioned its efficacy. Some raised concerns that service-learning might "dumb down" the curriculum. At Tufts, whose academic fortunes had clearly risen throughout the previous decade, innovative pedagogies were treated with particular suspicion. There were fears that widespread use of service-learning might be viewed as an emphasis unbefitting an elite research university. As one administrator observed: "For an aspiring institution, rigor is particularly important."

Institutions used several common strategies to promote service-learning. For example, the use of guest lecturers can enrich discussions by providing information about how other institutions have approached this task. A guest speaker can lend legitimacy to the project either symbolically (if he or she comes from a peer institution or better yet, an aspirational one) or by countering arguments with the growing body of research in civic engagement and service-learning. Specific workshops may be used to convey specific kinds of information, such as "service-learning 101." Public meetings also serve another important function. As the process moves from conveying an idea to jointly constructing a rationale for how it might be expressed in that specific context, public meetings are an occasion for proponents of the innovation to identify one another as a guiding coalition (Kotter 1996) and perhaps to reveal points of resistance. Over time, the conveyance of information shifts. For example, members of the organization will have marched sufficiently up the learning curve such that "how to" courses become less necessary or are targeted at certain groups, such as new students or faculty members. A different kind of communication occurs once the behavior becomes an institutional norm (as we discuss momentarily).

Encouraging the Behavior

Philosophical discussions of "service" and "social responsibility" are vitally important starting points. However, such lofty ideas must ultimately find expression in institutional behaviors and priorities. A primary goal must be supporting those *already* engaged in the behavior. Such individuals are an important resource about how service-learning fits into the organizational context, pedagogically (How does service-learning improve teaching and learning in my class?) and practically (What kinds of projects seem to suit our students best?). They can also be forceful advocates for this work. These institutions have also used a variety of strategies to encourage other individuals to move from thought to action.

First, they can lower the bar of participation by providing an easy way for those who show some interest to participate. Many individuals are initially unwilling to expend effort on something new because the activity doesn't yet seem important enough to supplant other professional tasks. Institutions can provide one-time service activities for students and often for faculty members. President Harris of Widener closed the university last year for Martin Luther King, Jr. Day and participated along with many faculty and staff members in planned service activities throughout the city of Chester.

Second, those willing to become more actively involved must also be encouraged and supported. Logistical support is vitally important for the success of service-learning initiatives. Swarthmore's director explains:

> If a faculty member wants to create a course, they can come here and we talk with them about what's involved and we create the community experience opportunity right down to arranging meeting with community partners—the professor is working with them from the beginning—what's best for the community *and* for the students? Eventually the professor will get a folder with the placements, [a staff member] will go to the class and talk to them about [the community] and if there's special training we provide that—talk to the students about how to enter a community, what is being arrogant and what is not. We can even help with reflection.

For decades Swarthmore offered little in the way of tangible support to faculty or students engaged in the community, with the exception of making grants to students for summer projects. In the 1980s, the Swarthmore Foundation was established and it began making more significant grants to faculty, staff, and students for community-based projects. In 1989, a "Green" Dean (recent alumnus/a) was hired to help students organize service activities. During President Bloom's administration, there were many conversations about how civic engagement efforts might be supported, spurred by Eugene Lang, chair of the board and a benefactor of the college. Early discussions about where to place the office led to the growing conviction that embedding it in any office would fail to provide a large enough umbrella for all the civic engagement activities (including service-learning). Such evolving support helps individuals to participate in service-learning in substantive and meaningful ways, which can lead to a greater sense of commitment on their part. Indeed, the experiences of these four institutions suggest that there is a close relationship between structural support and ideological support. Allocation of resources both symbolizes and is an essential element in producing support.

Third, institutions can encourage and reward individuals whose work supports the wider institutional effort. Last year Widener admitted its first group of Presidential Scholars. This group receives an education stipend, meets regularly as a cohort (often with the president in attendance), and, with the support of the special assistant to the president for Community Engagement, they are encouraged to develop a service project (Presidential Scholars are expected to perform 150 hours of community service each year). Swarthmore has for many years brought in a distinguished visiting professor whose work exemplifies community activism and engagement. Widener and Tufts are providing a course release to a group of tenured faculty members each year to enable them to develop new service-learning courses. Penn awards summer grants to faculty for curricular development as well.

The process of creating an environment conducive to the new behavior may require difficult decision-making. At one institution, it became evident that the provost, who had served the institution for some time, opposed the initiative. The provost had served the institution well for some time and had taken the lead on several curricular change efforts. He also enjoyed support from a group of influential faculty leaders. The incompatibility of the provost's curricular priorities and those of the service-learning proponents eventually led to the president asking him to resign.

Toward Normative Consensus

The position of any program or policy is ultimately tenuous—most can readily be wiped away at the whim of a new president. Institutionalization is more than a collection of administrative and curricular initiatives. As Goodman and Dean put it, it requires a normative consensus that the institutional behavior is not only acceptable but, in fact, it is an expression of the organization's core purpose. Some of this ideological groundwork is laid as people begin experiencing service-learning for themselves (or see how it adds value to their colleague's work). However, the perhaps unwelcome news for proponents of service-learning and civic engagement is that a true normative consensus must be formed one person at a time. Broad-based institutional efforts can bring the issue to the forefront and make it the subject of intense debate and conversation but ultimately individuals must conclude for themselves whether and how this work fits into their lives as scholars, teachers, administrators, and students. The good news is that if many individuals engage in this work, and normative consensus is achieved, it represents a powerful social contract among various constituencies that makes reversing the change extremely unlikely.

Ultimately, civic engagement and service-learning must be found to address important institutional imperatives: What is our unique institutional mission? What is the responsibility of our institution as part of this community? Who should we choose to join us in our effort—that is, which students should we admit and what faculty members should we hire? What should we expect of junior faculty members when they come up for promotion and tenure or merit raises? Responding to such questions requires long, careful, spirited, and at times even contentious, discussion. This is when the ideological battle is joined in earnest. Some of this may begin at the institutional level, for example, with the drafting of a new mission statement. But lofty statements must be translated into the nuts-and-bolts of daily life. It is therefore vitally important that the initiative be tied to larger institutional discussion of strategy and purpose. At Swarthmore, the decision to create a new center was arrived at during the planning of a capital campaign. To further highlight its importance, Jennie Keith, who had just returned to the faculty after two terms as provost, agreed to direct the center. The center director reports directly to the president. Departments must also individually decide for themselves how civic engagement may inform their work. This is where sufficient institutional support and sufficient number of proponents turns the tide.

As described earlier, there are numerous reasons why civic engagement efforts are resisted. Some academic norms rise to the level of conviction for certain faculty members. Many faculty members feel that the purpose of colleges and universities is to train the minds of students and that actually attempting to spur any sense of moral agency is either useless or a specious exercise: "Instruct students but don't try to encourage them to *do* anything to improve the world with that knowledge." There are also competing norms that may need to be addressed. For example, Swarthmore's long-standing ethos of service and the college's pride in the remarkable community-based projects initiated and led by students raised concerns that a more formal initiative might, as one administrator so eloquently put it, "pin the butterfly." The institutional value ("our students can achieve remarkable things all on their own") had to be amended to encourage greater participation by even more students. On a more prosaic level, change survivors (Duck 2001) will tend to view any institutional change (including civic engagement) as just one more passing fad. At Widener, a few faculty members voice the fear that service-learning might become a requirement even though President Harris had explained in a "town meeting" that service-learning is but one way of contributing to a civic engagement mission. These individuals may, however, over time be won over with adequate evidence and sustained effort (Hartley 2003).

Consensus building is no easy task and it requires the involvement of many, many groups. Administrative functions (e.g., student affairs), with the strong support of a president, can readily change their policies. Other administrative functions (e.g., admission policies) require the advice and consent of the faculty. Certainly any curricular initiative entails securing faculty support. But faculty members need to be convinced and to this end, senior faculty members are in the best position to advocate forcefully for an initiative. Further, they are immune from retribution if the department ultimately concludes that such activities are faddish or otherwise inappropriate. Junior faculty members are in a far more tenuous position. (No surprise that Widener selected tenured faculty members for its service grants and Tufts faculty fellows are full professors).

CONCLUSION

For lasting change to occur, proponents of service-learning must pay close attention to both structural and ideological elements, which, as table 13.1 demonstrates, differ considerably from one stage of institutionalization to another. Further, service-learning may become embedded through a succession of efforts with some schools or programs serving as a vanguard and others joining later. Perhaps the most important implication of these findings is that

Table 13.1   Examples of structural and ideological elements

Structural elements	Ideological elements
Pre-change	
• Pioneer's fund of knowledge about the pedagogy and the community	• Pioneers' commitment to service-learning and their ability articulate how it fits into their work
	• The institution's history (or founding mission) may support ideas such as a service or responsibility to community
Introducing the idea	
• Meetings: A way to gauge support or resistance and to begin identifying member of a coalition	• President's speeches articulating a new direction for the institution
• Drafting a mission statement: establishing institutional priorities and goals	• Meetings: A way to discuss and debate the efficacy of the initiative, to identify shared values/priorities, and to surface objections
• Workshops to teach people how to engage in service-learning	• Drafting a mission statement: Jointly articulating a rationale of the value of the initiative
Encouraging the behavior	
• Allocating resource for logistical support (e.g., staff support)	• Making symbolic policy changes (e.g., closing Widener for a day or service on MLK, Jr. Day)
• Creating "low-cost" ways to try the behavior	• "Low-cost" service opportunities allow people to begin to experience the activity for themselves
• Offering logistical support for those who want to become more involved.	• Allocating resources to support those who want to become more involved enables people to see the value in it and to incorporate it more fully into their work
• Allocating resources to support more sustained service-learning efforts by faculty/staff/students who want to become more involved.	• Allocating significant resources signals that the idea is no "fad"
	• Rewarding those already involved
Toward normative consensus	
• Linking the idea to institutional planning processes (e.g., strategic planning, capital campaign planning)	• Addressing competing values (e.g., the fear of discouraging student ownership and initiative, "pinning the butterfly")
• Creating new structures that signal institutional support at the highest level (e.g., The Swarthmore board's new standing committee on civic engagement)	• New structures at the highest level signal a lasting institutional commitment to this value

institutionalization requires continual cultivation and tending. None of the individuals we spoke with at these institutions feels any sense of complacency. Most were able to point to members of an "old guard"—faculty members for whom new modes of scholarship and teaching are somewhat suspect— whose support was unlikely and whose influence needed to be taken into account. As one director put it, "There are some faculty for whom we're simply irrelevant and that's just the way it is." Success lies in a slow and deliberate cultivation of supporters who can attest to service-learning's efficacy.

Swarthmore and Widener are beginning this process. Swarthmore's director explains:

> We're still moving out into groups of faculty that very much believe in what we're doing and wanted to do it but couldn't because they didn't have the resources. My hope is that as they talk about their experiences, we'll move out into the next ring of faculty members who haven't thought of this.

The more-established program at Penn and Tufts are already meeting the challenge of reaching into that "next ring" of faculty. However, the areas of scholarship and inclinations of these individuals in some cases make any connection to a broader civic engagement effort a more tenuous one than the early adopters. Each program faces considerable challenges. Hal Lawson (2002), in describing the ideal engaged campus notes: "All engaged universities are still evolving. All remain 'works in progress' " (p. 91).

The analysis presented here suggests that proponents of service-learning and civic engagement who seek institutionalization ought to approach their task like leaders of a grassroots movement—both structural and ideological issues must be addressed. It is insufficient to allocate institutional resources and staff lines for an innovation that has not gained widespread legitimacy within the organization. To do so is to build a Field of Dreams and hope "they" will come. Proponents must also seek to understand prevailing institutional norms and values: Is "service" or "responsibility to community" an important historic theme for the institution? If there are service-learning pioneers, how do they explain how their work fulfills their obligation to the institution and are there opportunities for larger conversation and dialogue? Conversely, an institution that eloquently redrafts its mission statement and publicly touts a renewed commitment to service but assumes that faculty and staff will carry the burden of sustaining the effort because of the intrinsic worth of the project has built a Potempkin Village—an elaborate façade that, while impressive, and apt to fool an outsider for a short time, will ultimately produce quite negligible results.

True institutionalization requires radical restructuring, the realigning of all the resources of the institution (structural and ideological) to a new and, in the current environment, somewhat contrarian purpose. Even at those institutions that have attempted to do both, significant challenges remain. It may be that total institutionalization by any institution is impossible in the current climate. Too many external factors (e.g., disciplinary expectations, institutional competition, commodification) hamper it. Perhaps our best hope is for individual institutions to continue to struggle but also for them to recognize that they are part of a larger movement that must challenge the norms of the entire academy to ultimately achieve complete success at the local level.

NOTES

1. See www.compact.org for additional information.
2. For additional detail, see www.upenn.edu/ccp.

3. http://uccps.tufts.edu/05_Faculty/faculty/html.
4. A complete discussion of this process can be found in Julie Reuben's book, *The Making of the Modern University: Intellectual Transformation and the Marginalization of Morality*. Chicago: University of Chicago Press, 1996.

REFERENCES

Benson, L. & Harkavy, I. (October 6–7, 2002). *Truly engaged and truly democratic cosmopolitan civic universities, community schools, and development of the democratic good of society in the 21st century*. Paper presented at the Seminar on the Research University as Local Citizen, University of California, San Diego.

Benson, L., Harkavy, I. & Hartley, M. (2005). Problem-solving service-learning in university-assisted schools as one practical means to develop democratic schools, democratic universities, and a democratic good society. In T. Chambers, J. Burkhardt, & A. Kezar (Eds.), *Higher education for the public good: Emerging voices from a national movement*. San Francisco: Jossey-Bass.

Boyer, E. (1990). *Scholarship reconsidered*. Princeton, NJ: The Carnegie Foundation for the Advancement of Teaching.

Diani, M. (1992). The concept of social movement. *The Sociological Review*: 1–25.

Duck, J. D. (2001). *The change monster: The human forces that fuel or foil corporate transformation and change*. New York, NY: Crown Business.

Glesne, C. & Peshkin, A. (1992). *Becoming qualitative researchers: An introduction*. New York: Longman.

Goodman, P. S. & Dean, J. W. (1982). Creating long-term organizational change. In P. S. Goodman (Ed.), *Change in organizations*. San Francisco, CA: Jossey-Bass.

Hartley, M. (2003). "There is no way without a because": Revitalization of purpose at three liberal arts colleges. *The Review of Higher Education*, 27(1): 75–102.

Hartley, M. & Hollander, E. (2005). The elusive ideal: Civic learning and higher education. In S. Fuhrman & M. Lazerson (Eds.), *Institutions of democracy: Public schools essay volume*. Oxford: Oxford University Press.

Hirschhorn, L. & May, L. (2000). The campaign approach to change: Targeting the university's scarcest resources. *Change*, 32(May/June): 30–37.

Hoffer, E. (1951). *The true believer: Thoughts on the nature of mass movements*. New York: Harper and Brothers.

Holland, B. (1997). Analyzing institutional commitment to service: A model of key organizational factors. *Michigan Journal of Community Service-learning* (Fall): 30–41.

Hollander, E. & Hartley, M. (2000). Civic renewal in higher education: The state of the movement and the need for a national network. In T. Ehrlich (Ed.), *Civic responsibility and higher education*. Phoenix, AZ: Orynx Press.

Kanter, R. M. (1983). *The change masters*. New York: Simon and Schuster.

Kotter, J. (1996). *Leading change*. Boston, MA: Harvard Business School Press.

Lawson, H. A. (2002). Beyond community involvement and service-learning to engaged universities. *Universities and Community Schools*, 7(1–2): 79–94.

Maurasse, D. (2001). *Beyond the campus: How colleges and universities form partnerships with their communities*. New York, NY: Routledge.

Maxwell, J. A. (1996). *Qualitative research design: An interactive approach*. Thousand Oaks, CA: Sage.

Ostrander, S. A. (2004). Democracy, civic participation, and the university: A comparative study of civic engagement on five campuses. *Nonprofit and Voluntary Sector Quarterly*, 33(1): 74–93.

Patton, M. Q. (1990). *Qualitative evaluation and research methods* (2nd ed.). Newbury Park, CA: Sage.

Rossman, G. B. & Rallis, S. F. (1997). *Learning in the field: An introduction to qualitative research.* Draft.

Schein, E. H. (1985). *Organizational culture and leadership* (2nd ed.). San Francisco: Jossey-Bass.

Singleton, S., Burack, C. A. & Hirsch, D. J. (1997). *Faculty service enclaves: A summary report.* Boston, MA: University of Massachusetts, Boston.

Stanton, T. K., Giles, D. E., Jr. & Cruz, N. I. (1999). *Service-learning: A movement's pioneers reflect on its origins, practice and future.* San Francisco, CA: Jossey-Bass.

Ward, K. (1996). Service-learning and student volunteerism: Reflections on institutional commitment. *Michigan Journal of Community Service-learning,* 3: 55–65.

Weick, K. E. (1976). Educational organizations as loosely coupled systems. *Administrative Science Quarterly,* 21: 1–19.

INDEX

Printed in the United States
206727BV00003B/5/A

9 781403 968777